Worked Examples in the Kinetics and Thermodynamics of Phase Transformations

E A Wilson

Published by

The Institution of Metallurgists
Northway House, High Road
Whetstone, London N20 9LW

Printed by The Chameleon Press Limited.
5-25 Burr Road, Wandsworth, London SW18 4SG.

Preface

It is a pleasure to acknowledge the excellent teaching of kinetics of phase transformations by Professor W S Owen and Professor J Burke which first inspired the author's interest in phase transformations in the early sixties.

To the author's knowledge there are now no cheap student's texts available in this country on the topic of kinetics of phase transformation so it was decided to include some derivations of the classical equations in kinetics in the present monograph along with specific worked examples.

Since the appearance of Burke and Fine there has been a re-emphasis of the importance of the Zeldovitch non-equilibrium factor together with a possible explanation of the nature of incubation periods. Aaronson and Lee's treatment of these topics is now out of print and therefore a summary of the derivation of the Zeldovitch non-equilibrium factor is included in the monograph.

The author has drawn heavily on the books by Christian, Burke, Fine and Aaronson and Lee for which he is grateful. The monograph is not intended to be a full quantitative treatment of the kinetics of phase transformations many of the more complete and sophisticated derivations being outside the syllabus of most undergraduate students. To illustrate the limitation of most theoretical treatments, the data for the worked examples has been taken from actual experimental results in the literature.

Clearly any undergraduate teaching should include topics not included in the present monograph, particularly some of the more qualitative descriptions of phase transformations as yet not amenable to quantitative treatments. Nevertheless it is hoped that readers, particularly students, will find the present text useful.

In conclusion I would like to acknowledge useful discussions with Dr R Smith and Dr B Hattersley of the Polytechnic, Professor Greenwood, Sheffield University, Dr C N Reid, Open University and Dr T Gladman of BSC

Swinden Laboratories for various parts of the text. The encouragement and patience of Dr F B Pickering was helpful in producing the text.
Mr M Motley kindly proof read the whole of the manuscript but the author of course takes full responsibility for any errors or mistakes in the monograph.
Last of all, but by no means least, I would like to thank Jennifer Senior for her skill and patience in typing the monograph and Jack Evans for drawing the figures.

Introduction

Properties of materials depend to a large extent on their microstructure.

In metals these microstructures are mainly developed as a result of phase changes both on solidification and phase changes at lower temperatures in the solid state. It is therefore important to study why these phase changes occur and how they occur.

In the main, the reason why phase changes can occur can be obtained from a proper study of the phase diagram, while a knowledge of how transformations occur can be obtained from a study of the kinetics of phase transformations.

This monograph attempts to describe some of the simple quantitative treatments of phase transformations. However the text is not complete by any means and many of the more qualitative treatments of phase transformations should also be studied by the student. For example the important topic of growth by movement of ledges is not covered, nor the bainite transformation, or crystallographic theories of martensite formation.

Many of the topics covered are taken from ferrous metallurgy. This is partly a reflection of the author's research and teaching experience, but is also due to the fact that solid state transformations are used in steels to produce a required microstructure by heat treatment, and steel is one of the cheapest engineering materials available. Consequently phase transformations have been more thoroughly studied in steels than other metals.

A chapter on recovery, recrystallisation and grain growth is included in the monograph, since although these processes are not strictly speaking phase transformations they do involve a change in state of the metal and are therefore influenced by time and temperature in a similar manner to phase changes proper.

An attempt has been made to keep the mathematics simple, and a summary of the formulae and relationships used in the monograph is given on page (v) Similarly a summary of the relevant thermodynamic relationships is given on page (vi).

Problems in the Kinetics and Thermodynamics of Phase Transformations

Summary of essential mathematics used in monograph

In the right angled triangle shown, by Pythagoras

$$C^2 = a^2 + b^2$$

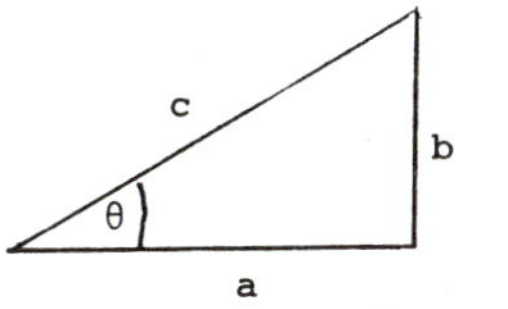

$\sin\ \theta = \frac{b}{c}$; $\cos\ \theta = \frac{a}{c}$

$$\therefore \quad \sin^2\theta + \cos^2\theta = 1$$

$\sin\theta$ and $\cos\theta$ vary with θ as shown.

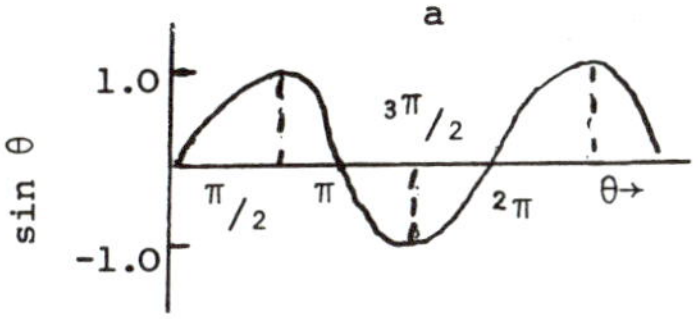

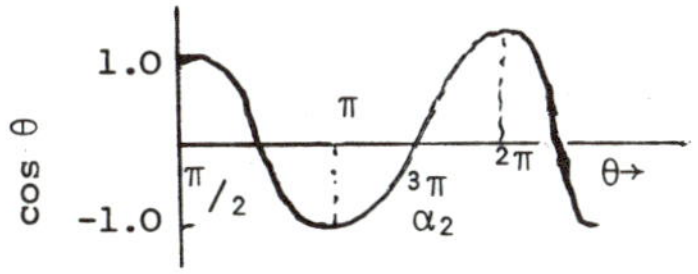

For a quadratic

$$ax^2 + bx + c = 0$$

$$x + \frac{-b \pm \sqrt{b^2 - 4ac}}{2a}$$

For real roots

$$b^2 > 4ac$$

$$\frac{d}{dx}(ax^n) = anx^{n-1}$$

$$\frac{d}{dx}(\sin x) = \cos x$$

$$\frac{d}{dx}(\cos x) = -\sin x$$

$$\frac{d}{dx}(\ln x) = \frac{1}{x}$$

$$\frac{d}{dx}(ae^{kx}) = ake^{kx}$$

$$\int ax^n dx = \frac{ax^{n+1}}{n+1} + \text{Constant}$$

$$\frac{d}{dx}(uvw) = \left(\frac{1}{u}\frac{du}{dx} + \frac{1}{v}\frac{dv}{dx} + \frac{1}{w}\frac{dw}{dx}\right)(uvw)$$

$$\int \frac{a\,dx}{x} = a\ln x + \text{Constant}$$

$$\int \cos x\,dx = \sin x + \text{Constant}$$

$$\int \sin x\,dx = -\cos x + \text{Constant}$$

If $y = e^x = \exp(x)$ then $x = \ln y$

$\ln 1 = 0$

$$\ln\frac{1}{x} = \ln 1 - \ln x = -\ln x$$

$$e^x = 1 + \frac{x}{1!} + \frac{x^2}{2!} + \frac{x^3}{3!} + \frac{x^4}{4!}$$

Error Function $\phi = \frac{2}{\sqrt{\pi}} \int_0^{\lambda} \exp(-\lambda^2)\, d\lambda$

$\phi = 0$ when $\lambda = 0$

$\phi = 1.0$ when $\lambda = \infty$

$= 1.0$ when $\lambda = -\infty$

Taylor series for expansions of f(x) about a value $x = x_o$

$$f(x) = f(x_o) + \frac{(x - x_o)}{1!} f'(x_o) + \frac{(x - x_o)^2}{2!} f''(x_o) + \ldots\ldots\ldots$$

<u>Summary of Chemical Thermodynamics used in monograph</u>

Gibbs Free Energy $G = H - TS$

For the difference in Gibbs Free Energy between two phases α and γ

$$\Delta G^{\gamma\to\alpha} = \Delta H^{\gamma\to\alpha} - T\Delta S^{\gamma\to\alpha}$$

If H and S are referred to a reference state of 298K then

$$\Delta H_{298}^{\gamma\to\alpha} = \int_{298}^{T} \Delta C_P^{\gamma\to\alpha} dT \text{ where } \Delta C_P^{\gamma\to\alpha} = C_P^{\alpha} - C_P^{\gamma}$$

$$\Delta S_{298}^{\gamma\to\alpha} = \int_{298}^{T} \frac{\Delta C_p^{\gamma\to\alpha}}{T} dT$$

For a reaction involving an equilibrium constant K

$$\Delta G = kT\ln K$$

Boltzmann statistics gives us that the number of particles n in a system having an energy Q is given by

$$n = N \exp(-\frac{Q}{kT})$$

where N = total number of particles.

Gibbs Duhem for a two component system containing atom fractions N_1 and N_2 of components (1) and (2).

$$N_1 \frac{d^2G_1}{dN_1{}^2} + N_2 \frac{d^2G_2}{dN_2{}^2} = 0$$

1. Solid Solubility and the Solubility Product

Consider the solid solubility of the phase A_mB_n in the solid solution phase α in the pseudo-binary phase diagram shown in figure 1.1.

At the phase boundary at a particular temperature T kelvin, the following reaction can occur:

$$m[A] + n[B] \rightarrow A_mB_n \qquad 1.1$$

In α solid solution

where [A] = solubility limit of A in α expressed as mole fraction

[B] = solubility limit of B in α expressed as mole fraction.

The equilibrium constant K for this reaction (1.1) is given by:

$$K = \frac{[a_{A_mB_n}]}{[a_A]^m[a_B]^n} \qquad 1.2$$

$$\Delta G^o = -RT\ln K \qquad 1.3$$

where $a_{A_mB_n}$ = activity of compound A_mB_n

a_A = activity of element A in α

a_B = activity of element B in α

ΔG^o = Free energy of formation of A_mB_n from the solid solution α

For a pure compound $a_{A_mB_n}$ = 1 and for dilute solutions we can assume ideal behaviour

$$a_A = [A] \text{ and } a_B = [B] \qquad 1.4$$

Hence we obtain from equations 1.2, 1.3 and 1.4

$$\Delta G^o = RT\ln[A]^m[B]^n. \qquad 1.5$$

$$\text{Now } \Delta G^o = \Delta H^o - T\Delta S^o \qquad 1.6$$

where ΔH^o = Heat of formation of A_mB_n from the solid solution α)
ΔS^o = Entropy of formation of A_mB_n from the solid solution α.) 1.7

Assuming ΔH and ΔS do not vary with temperature we obtain from equations 1.5 and 1.6

$$\ln[A]^m[B]^n = \frac{\Delta H^o}{R} \times \frac{1}{T} - \frac{\Delta S^o}{R} \qquad 1.8$$

This equation (1.8) can be written as

$$\log K_s = \log[\%A]^m[\%B]^n = \frac{C}{T} + B \qquad 1.9$$

Solubility data are often expressed in the form of equation 1.9 where,

K_s = Solubility product for A_mB_n = $[\%A]^m[\%B]^n$.

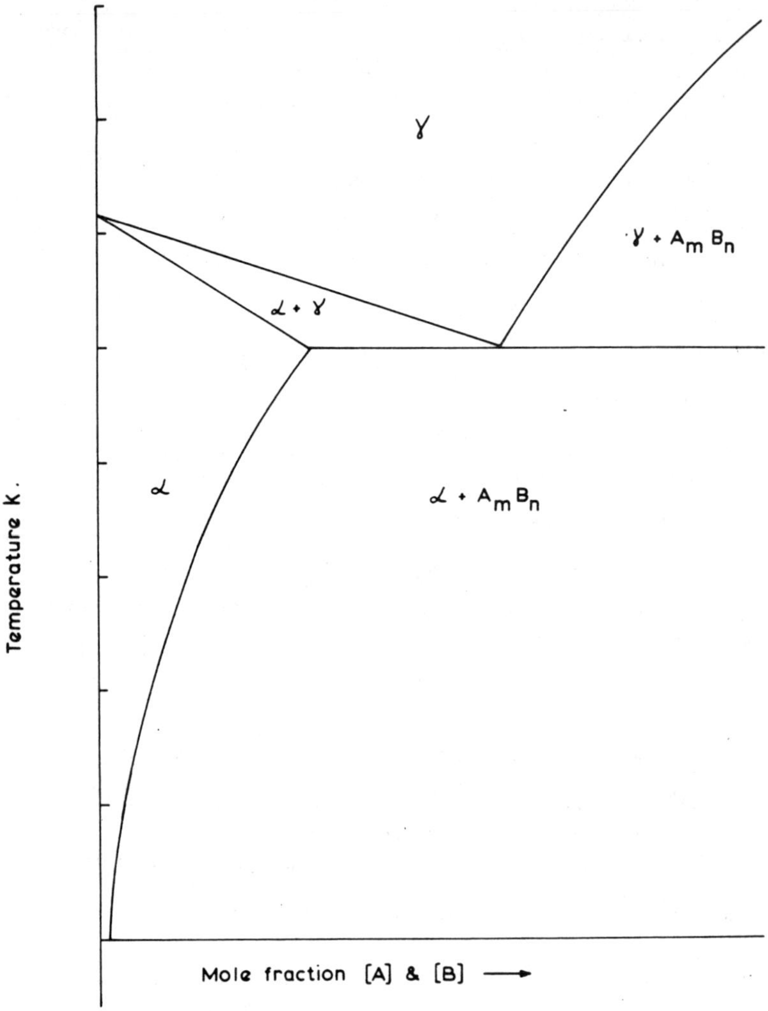

FIG. 1.1 Schematic diagram showing the variation of solid solubility of $A_m B_n$ with temperature

Problem 1.1 Q.5 BSc (Hons) Final Year Metallurgy, Physical Metallurgy, June 1976, Sheffield Polytechnic.

The following data shows the variation of solid solubility of copper in ferrite:

Temperature ^{o}C	841	801	759	728	698
Solid solubility, atomic %	1.79	1.22	0.86	0.61	0.44

Calculate the heat of formation of copper from the solid solution and the entropy of formation of copper from the solid solution.

What is the physical interpretation of these quantities?

Comment on the accuracy of your determined values of these quantities.

Gas content $R = 8.314$ J/mol K

The following reaction occurs at the solubility limit C (mole fraction) at temperature T kelvin.

$$\alpha \longrightarrow Cu$$

Cu in solid solution	Essentially pure Cu

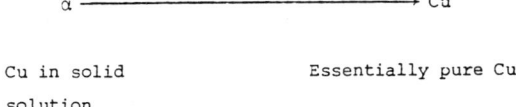

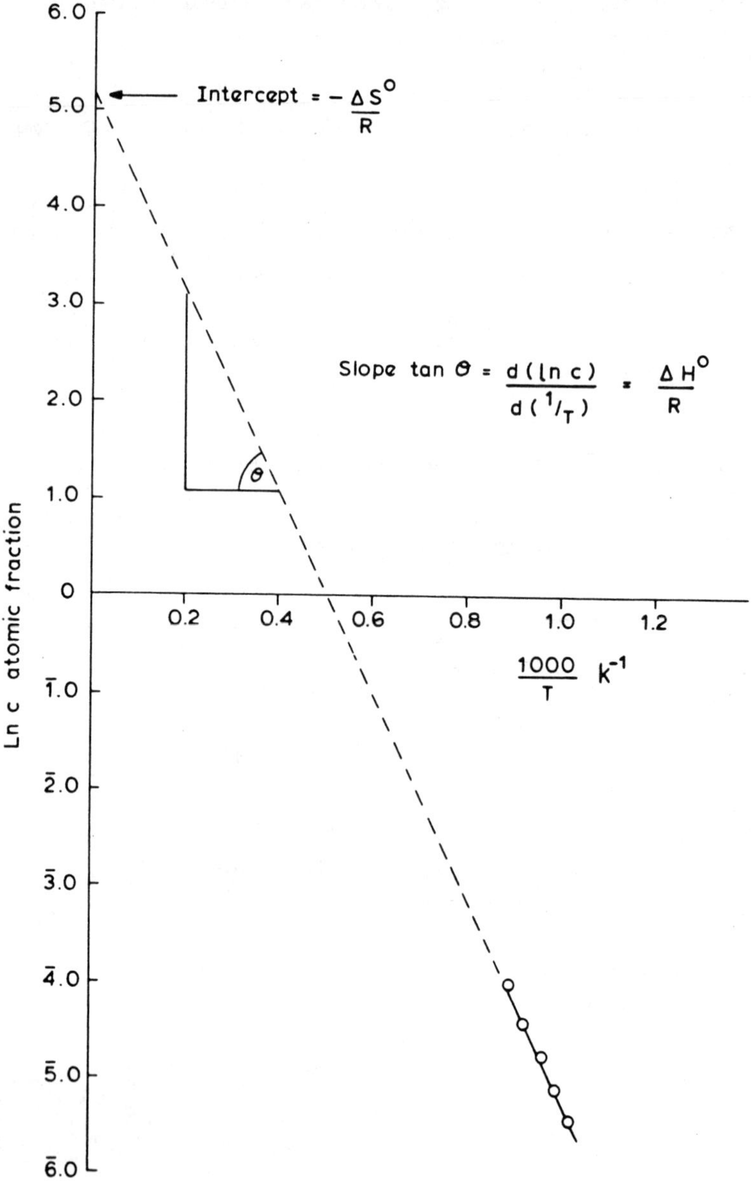

FIG. 1.2 Solubility of Cu in α - Fe.

The rate constant K for this reaction is given by, assuming ideal solid solution

$$K = \frac{[a_{Cu}]}{[a^{\alpha}_{Cu}]} = \frac{1}{C} \quad \ldots\ldots \quad 1.11$$

$$\Delta G^{o} = -RT \ln K \quad \ldots\ldots \quad 1.12$$

$$= \Delta H^{o} - T\Delta S^{o} \quad \ldots\ldots \quad 1.13$$

where a^{α}_{Cu} = activity of Cu in αFe = C, if ideal.

a_{Cu} = activity of Cu out of solid solution = 1, if ideal.

ΔG^{o} = Free energy of formation of Cu from α/mol of Cu

ΔH^{o} = Heat of formation of Cu from α/mole of Cu

ΔS^{o} = Entropy of formation of Cu from α/mol of Cu

Hence combining equations 1.11, 1.12 and 1.13

$$\ln C = \frac{\Delta H^{o}}{R}\frac{1}{T} - \frac{\Delta S^{o}}{R}$$

$\therefore$ A plot of ln C versus 1/T is linear of slope $(\frac{\Delta H^{o}}{R})$ and intercept $(-\frac{\Delta S^{o}}{R})$

Referring to figure 1.2,

$$\text{Intercept} = -\frac{\Delta S^{o}}{R} = 5.15$$

$\therefore$ $\Delta S^{o} = -8.314 \times 5.15 = \underline{\underline{-43 \text{ J/mol K}}}$

$$\text{Slope} = \frac{\Delta H^{o}}{R} = -\frac{(3.1 - 1.075)}{0.2 \times 10^{-3}} = -10.125 \times 10^{3}$$

$\therefore$ $\Delta H^{o} = -8.314 \times 10.125 = \underline{\underline{-84 \text{ kJ/mol}}}$

Note C is plotted as atom or mole fraction not atomic %.

The fact that ΔH^{o} is negative means that there is a negative heat of formation or positive heat of solution. This means there is a preference for Fe-Fe and Cu-Cu bonds rather than Fe-Cu bonds. This is to be expected in view of the small solid solubility.

ΔS^{o} measures the configuration /entropy which will be small when there is limited solubility.

Because of the long extrapolation involved ΔS^{o} will not be very accurate; ΔH^{o} obtained from the slope of the line will be more accurate.

Problem 1.2 Q.4 Institution of Metallurgists examinations, AIM Paper 1, Physical Metallurgy A, July 1969.

In the simple eutectic system for copper and silver the limiting solubility of silver in copper at various temperatures is listed as:

Temperature ^{o}C	597	679	769
Atomic % of silver	1.4	2.4	4.15

Estimate the heat of formation of silver, and the entropy of formation of silver from the solid solution.
(Approximately two thirds of the marks were allocated to the calculation).

Problem 1.3 State Hume-Rothery's size factor rule governing the solid solubility of primary phases in binary alloys.

The solid solubility of Pb in Cu is 0.09 atomic per cent at 600°C.

Calculate the change in free energy associated with this solubility. You may assume that solid solution of Pb in Cu is thermodynamically ideal.

Show that the misfit strain determined from this energy obeys Hume-Rothery's size factor role.

Data:- Atomic diameter of Cu = 2.556 Å ; 1Å = 10^{-10}m
Bulk modulus of Pb, K = 50 kN/mm^2
Avogadro's number N_o = 6.0225 x 10^{23}
Gas constant R = 8.314 J/mol K.

"If the atomic diameters of solvent and solute differ by more than 15% then the size factor is unfavourable and solid solubility is very restricted, whilst when the atomic diameters are within this limit the size factor is favourable and considerable solid solubility may occur if other factors are favourable."

Proceeding as before in problem 1.1 we obtain at the solubility limit, C,

$$\Delta G^o = -RT \ln K = -RT \ln \frac{1}{C}$$

$$= -8.314 \times 873 \ln \frac{1}{9 \times 10^{-4}} = \underline{\underline{-55.9 \text{ kJ/mol}}}$$

We now have to obtain an expression for ΔG^o due to the linear dilatation Δ accompanying the presence of a Pb atom in the copper matrix.

We will assume that the copper matrix is rigid and that the size of the hole, in which we insert a Pb atom, r_o = 1.278Å . Further we assume that purely elastic deformation occurs to the Pb atom on insertion of the Pb atom into the hole radius r_o, giving a linear dilatation = Δ.

On insertion of the Pb atom, it will be subject to a uniform hydrostatic pressure building up from zero to a value P.

$\therefore$ work done/atom = ½PdV

where dV = volume change of Pb atom in insertion in hole of radius r_o.

$\therefore$ Work done/atom $= \frac{1}{2}$ PdV

$$= \frac{1}{2} K \frac{dV^2}{V} \qquad \text{where } K = \frac{P}{dV/V}$$

$$= \frac{1}{2} K \left(\frac{dV}{V}\right)^2 V = \frac{1}{2} K(3\Delta)^2 V.$$

Since, $V = \frac{4}{3} \pi r_o^{\,3}$

$dV = 4 \pi r_o^{\,2}\, dr$

$$\left.\right\} \quad \frac{dV}{V} = \frac{3dr}{r_o} = 3\Delta$$

$\therefore$ Work done/atom $= \frac{9}{2} KV\Delta^2 = 6K\pi r_o^{\,3}\Delta^2$

$\therefore$ Work done/mole $\Delta W = 6N_o \pi K r_o^{\,3} \Delta^2$

$$= 6 \times 6.0225 \times 10^{23} \times 5 \times 10^{10} \; (1.278 \times 10^{-10})^3 \pi \Delta^2$$

$$= \underline{1,847\Delta^2 \text{ kJ/mol}}$$

ΔW is the work done on insertion of the Pb atom, whereas ΔG^o is the free energy released on forming free lead from the solid solution.

$\therefore \Delta W = -\Delta G^o$

$\therefore \Delta^2 = \frac{5.59 \times 10^4}{1.847 \times 10^6} = 3.027 \times 10^{-2}$

$\therefore \Delta = 1.74 \times 10^{-1} = \underline{\underline{17.4\%}}$

The actual misfit calculated from atomic diameters of Pb and Cu atoms is 37%.

In this derivation, the misfit strain energy $\Delta W = 9/2KV\Delta^2 = 12GV\Delta^2$ [assuming Poisson's ratio $\nu = 1/3$; G = Shear modulus, and $K = \frac{2G(1 + \nu)}{3\ (1 - 2\nu)}$]. More realistic misfit strains are obtained if the misfit strain energy is taken as $6GV\Delta^2$(1.1, 1.2). This is the result for insertion of a rigid sphere in an elastic matrix.

As Darken and Gurry (1.2) point out "classical elastic theory can hardly be expected to apply for such a large displacement (∿ 15%) at the atomic scale"; nevertheless the theory provides "a very understandable basis for Hume-Rothery's empirical rule".

<u>Problem 1.4</u> An Fe-8% Mn alloy containing 0.008%N, 0.018%C, 0.13% soluble Ti and 0.07% soluble Al is ice brine quenched from 900°C. Make an estimate of the upper limit of nitrogen in solid solution in this alloy at room temperature after this heat treatment.

Data:

Element	Relative atomic mass
Fe	55.85
Mn	54.94
Ti	47.9
Al	26.97
C	12.01
N	14.008

In HSLA steels (1.3)

Solubility product at $900^{\circ}C$ for TiC = 2.97×10^{-3} (atomic %)^2

Solubility product at $900^{\circ}C$ for AlN = 3.16×10^{-4} (atomic %)^2

Solubility product at $900^{\circ}C$ for TiN = 1.46×10^{-4} (atomic %)^2

$$0.008 \text{ mass } \%N = \frac{\frac{0.008}{14.008} \times 100}{\frac{0.13}{47.9} + \frac{0.07}{26.97} + \frac{0.018}{12.01} + \frac{0.008}{14.008} + \frac{8}{54.94} + \frac{91.774}{55.885}}$$

$$\simeq \frac{0.008}{14.008} \times 55.85^*$$

$$\simeq \underline{0.032 \text{ atomic } \%N}$$

$$0.018 \text{ mass } \%C \simeq \frac{0.018}{12.01} \times 55.85^*$$

$$\simeq \underline{0.084 \text{ atomic } \%C}$$

$$0.13 \text{ mass } \% \text{ Ti} \simeq \frac{0.13}{47.9} \times 55.85^*$$

$$\simeq \underline{0.152 \text{ atomic } \%}$$

$$0.07 \text{ mass } \% \text{ Al} \simeq \frac{0.07}{26.97} \times 55.85^*$$

$$\simeq \underline{0.145 \text{ atomic } \%}$$

A schematic diagram illustrating the solubility of TiC, TiN and AlN at $900^{\circ}C$ and the precipitation line for these phases in the alloy is given in figure 1.3.

We will assume that it is possible that all the C combines with Ti as TiC and that N is equally partitioned between the remaining Ti and Al.

* The approximation in fact only results in a difference at the fourth decimal place.

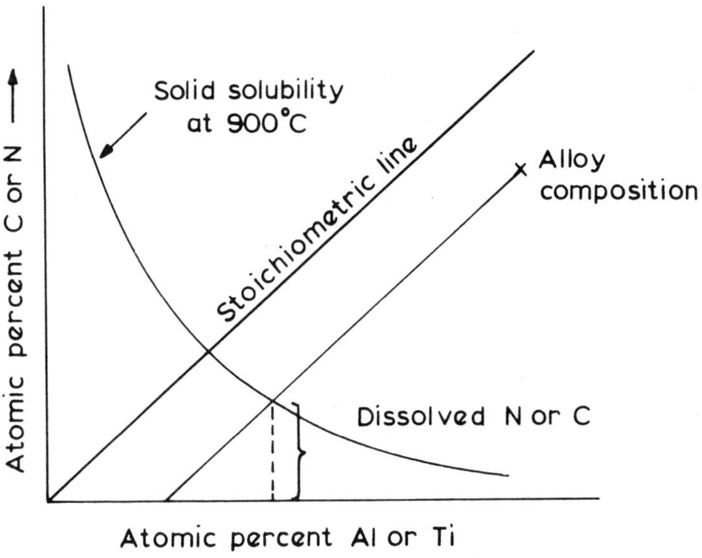

FIG. 1.3 Solubility of N and C in a stabilised steel.

0.084 a/o C requires 0.084 a/o Ti) ∴ excess Ti
for complete precipitation) = 0.152 - 0.084 - 0.016
0.032 a/o N requires (0.016 a/o Ti) = 0.052 a/o
(0.016 a/o Al) Total Al in alloy = 0.145 a/o
for complete precipitation) leaving 0.129 a/o excess Al

We will further assume that the excess Ti of 0.052 a/o is equally divided between carbon (0.026 a/o) and nitrogen (0.026 a/o).

The equation of line along which precipitation of TiN occurs is therefore given by

$$\text{a/o N} = \text{a/o Ti} - 0.026$$

The equation of the solubility curve at 900°C for TiN is given by

$$(\text{a/o N})(\text{a/o Ti}) = 1.46 \times 10^{-4}$$

∴ at intersection of precipitating line and solubility curve

$$(\text{a/o N})(\text{a/o N} + 0.026) = 1.46 \times 10^{-4}$$

ie $(\text{a/o N})^2 + 0.026\,(\text{a/o N}) - 1.46 \times 10^{-4} = 0.$

∴ a/o N in solid solution uncombined with Ti

$$= \frac{-0.026 \pm \sqrt{(0.026)^2 + (4 \times 1.46 \times 10^{-4}}}{2}$$

$$= 0.0047 \text{ atomic N}$$

$$= 0.0047 \times \frac{14.008}{55.85} = \underline{0.0012 \text{ mass \%N}}$$

The equation of the line along which precipitation of AlN occurs is given by:

$$\text{a/o N} = \text{a/o Al} - 0.129$$

The equation of the solubility curve at 900°C for AlN is given by

$$(\text{a/o N})(\text{a/O Al}) = 3.16 \times 10^{-4}$$

∴ at intersection of precipitating line and solubility curve

$$(\text{a/o N})(\text{a/o N} \times 0.129) = 3.16 \times 10^{-4}$$

i.e. $(\text{a/o N})^2 + 0.129(\text{a/o N}) - 3.16 \times 10^{-4} = 0$

∴ %N in solid solution uncombined with Al

$$= \frac{-0.129 \pm \sqrt{(0.129)^2 + (4 \times 3.16 \times 10^{-4}}}{2}$$

$$= 0.0024 \text{ atomic \%N}$$

$$= 0.0024 \times \frac{14.008}{55.85} = \underline{0.0006 \text{ mass \%N}}$$

$\therefore$ total N in solid solution

$= 0.0012 + 0.0006 = 0.0018$

$<$ 0.002 mass %N

Ellingham Diagrams

These are a useful concept to determine which sulphide, oxide, nitride or carbide will form in the presence of various alloying elements.

On an Ellingham Diagram (1.5) the energy of formation of the compound ΔG^o is plotted against temperature. To a first approximation these are straight lines since in the equation $\Delta G^o = \Delta H^o - T\Delta S^o$ ΔH^o and ΔS^o do not vary appreciably with temperature. However, at a change of state such as boiling point of the element (point B in the daigrams), the slope changes because disorder increases, i.e. entropy increases. (There is more disorder in a gas than in a solid).

Since these diagrams give the energy of formation of the various compounds, those with the largest negative value of ΔG^o (i.e. lowest position on the diagram) will be the most stable and likely to form in the presence of other alloying elements.

Thus for example in figure 1.4 (1.6), in the presence of manganese and sulphur, MnS is more likely to form in a steel than FeS. The hot shortness of steel due to the formation of FeS can therefore be avoided by the addition of Mn. It will also be seen that CaS and CeS are extremely stable and hence explain their use in inclusion shape control.

In figure 1.5 (1.6) it will be seen that Al_2O_3 is very stable and thus explains the use of aluminium as a de-oxidant in steels, while in figure 1.6 the stability of nitrides increases in the order Al,Ti and Zr. Thus in problem 1.4

there will actually be more partitioning of nitrogen towards Ti.

However, care should be used in predicting the effect of alloying element on compound formation in steels, since the activity of the alloying element changes on alloying with iron. This is illustrated by comparing figures 1.6 and 1.7. It will be seen that the slope of the lines for each compound increases, because although there is a strong attraction between the solute atoms and carbon, there is also some interaction between the solute atom and iron atoms. There is therefore, more randomness in the system Fe-X-C than X-C and entropy increases.

References

1.1 F.R.N. Nabarro. Proc. Roy Soc., 1940, 175A, p.519.

1.2 L.S.Darken and R.W.Gurry. "Physical Chemistry of Metals", McGraw-Hill, New York, 1953, p.78.

1.3 B.Aronsson. "Steel-Strengthening Mechanisms", Climax Molybdenum Symposium, Zurich, May 1969, pp.78-87.

1.4 T.Gladman, D. Dulieu and I.D. McIvor. "Micro-Alloying 75", Union Carbide Corporation, 1975, pp 26-48.

1.5 H.J.T. Ellingham. J. Soc. Chem. Ind. 1944, 63, p.125.

1.6 F.D.Richardson and J.H.E.Jeffes. J. Iron Steel Inst., 1952 171, pp.165-175.

1.7 F.D.Richardson. J. Iron and Steel Inst., 1953, 175 pp.33-51.

1.8 J.Pearson and Ursula J.C.Ende. J. Iron and Steel Inst. 1953, 175 pp.52-58.

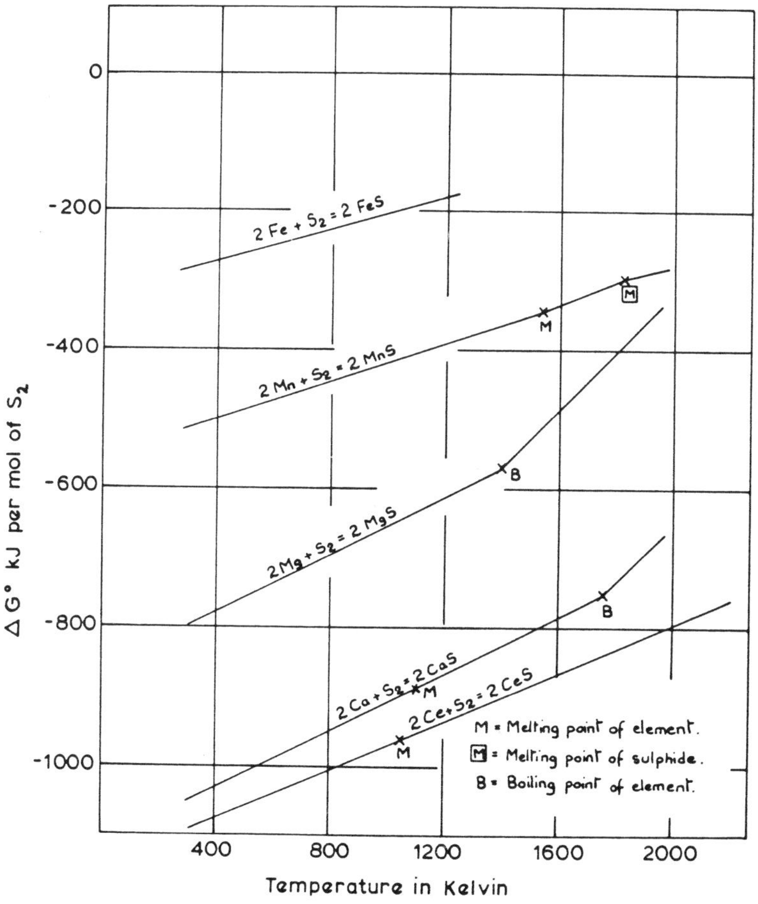

FIG. 1.4 ELLINGHAM DIAGRAM FOR STABILITY OF SULPHIDES.
(After Richardson)

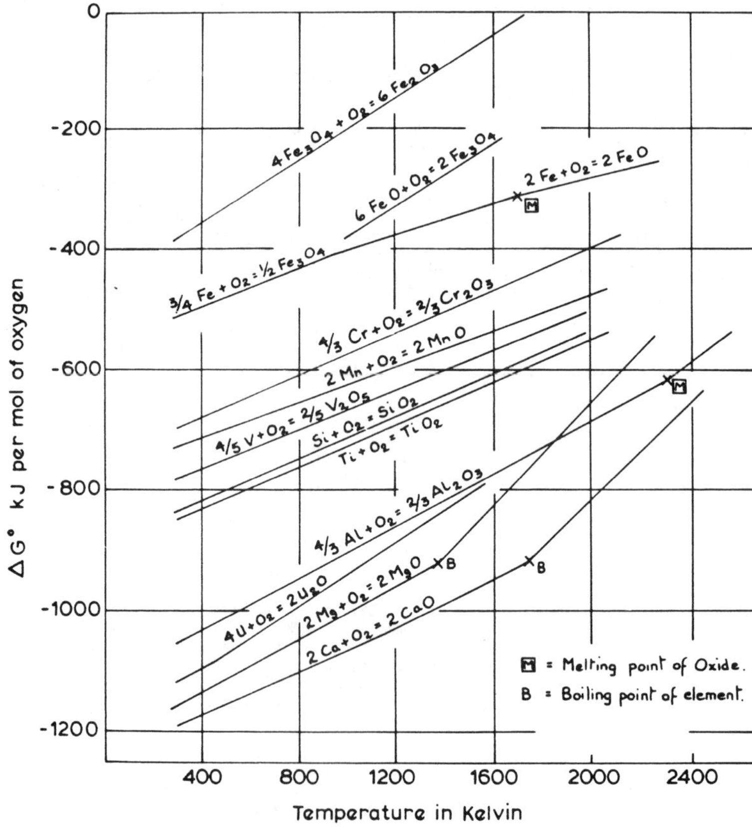

FIG. 1.5 ELLINGHAM DIAGRAM FOR STABILITY OF OXIDES (After Richardson)

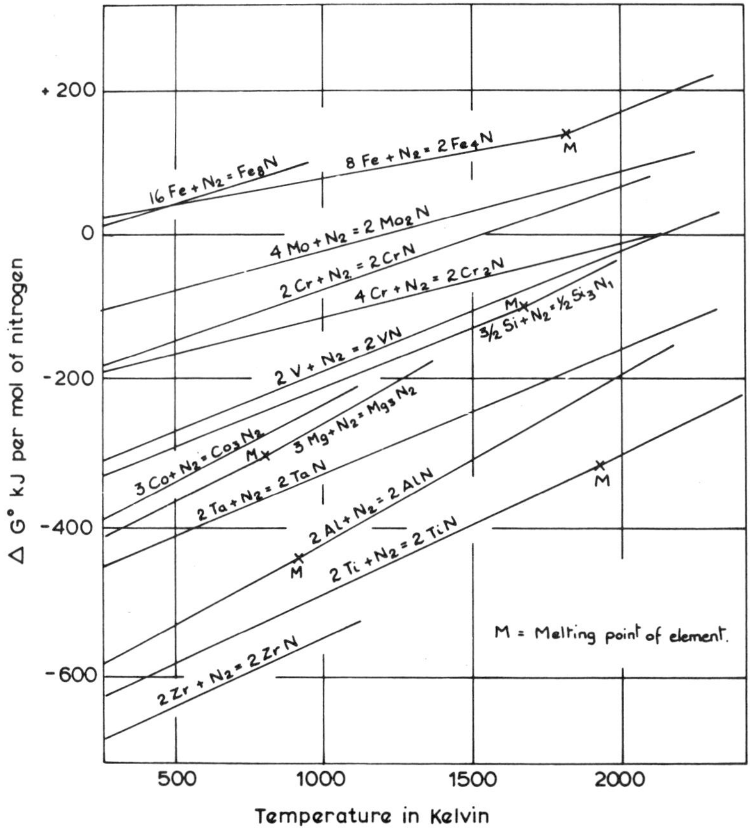

FIG. 1.6 ELLINGHAM DIAGRAM FOR STABILITY OF NITRIDES.
(After Pearson and Ende)

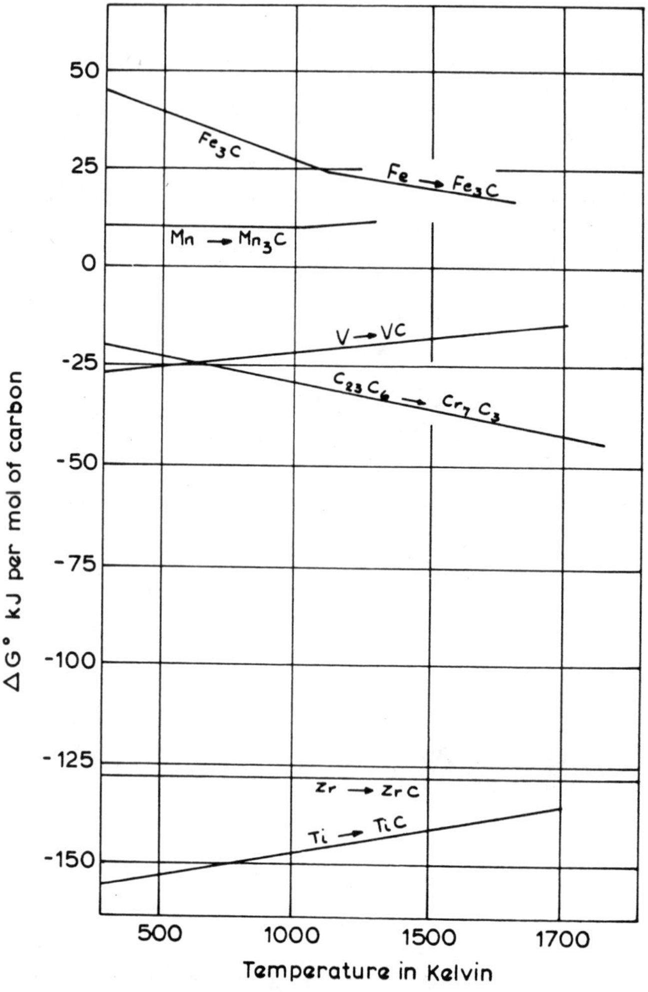

FIG. 1.7 ELLINGHAM DIAGRAM FOR STABILITY OF CARBIDES (After Richardson)

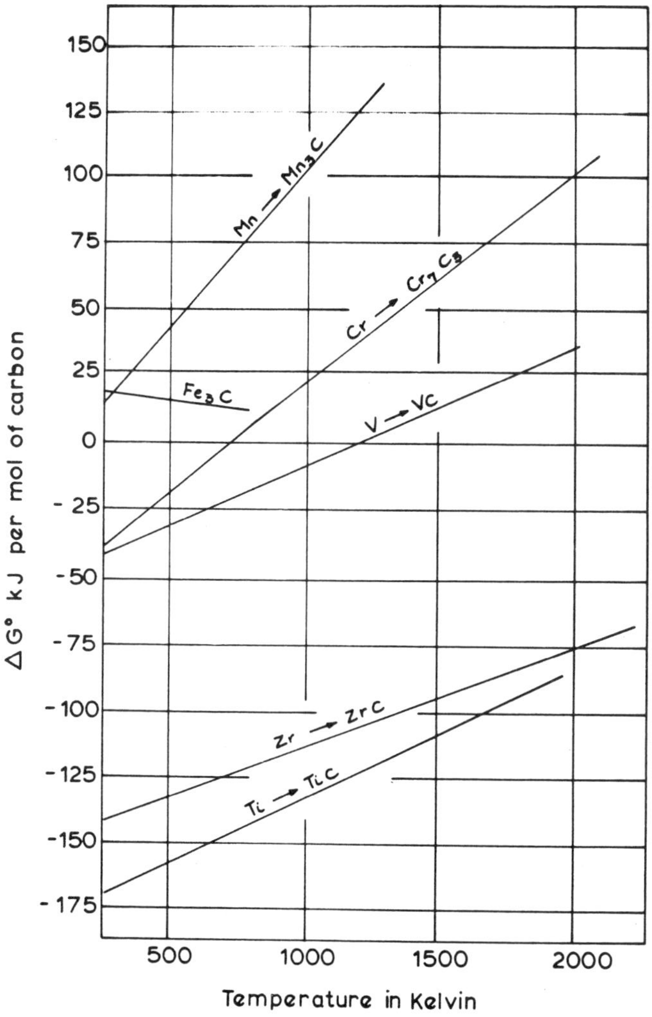

FIG. 1.8 ELLINGHAM DIAGRAM FOR STABILITY OF CARBIDES IN STEEL, OVERHEATED AT 1% ACTIVITY OF ALLOYING ELEMENT IN IRON. (After Richardson)

2. Chemical Driving Force for Transformations

2.1 Concept of T_o (2.1)

Consider the variation of Gibbs free energy/unit volume, G_v, with temperature for two phases α and γ; figure 2.1. At high temperatures G_v^{γ} is less than G_v^{α} and therefore γ is the equilibrium phase, while at low temperatures $G_v^{\alpha} < G_v^{\gamma}$ and α is the equilibrium phase. At some temperature T_oK, $G_v^{\alpha} = G_v^{\gamma}$ and therefore the phases α and γ are in equilibrium. Thus for example in pure iron $T_o = 910^{o}C = 1183K$ for equilibrium between ferrite α and austenite γ.

At some temperature T_1 below T_o, the chemical driving force for the reaction γ→α is given by

$$\Delta G_v = G^{PRODUCTS} - G^{REACTANTS}$$

$$= G_v^{\alpha} - G_v^{\gamma}$$

$$\equiv \text{-ve below } T_o.$$

In general the free energies curves for G_v^{α} and G_v^{γ} will not vary linearly with temperature, but a reasonable approximation particularly for small degrees of undercooling, is to assume that G_v^{α} and G_v^{γ} do vary linearly with temperature.

This means that in the following equation ΔH_v and ΔG_v do not vary with temperature:-

$$\Delta G_v = \Delta H_v - T\Delta S_v \quad \text{.......... 2.1}$$

Now at T_o, $\Delta G_v = 0$

$$\therefore \Delta H_v = T_o \Delta S_v \quad \text{.......... 2.2}$$

Substituting equation 2.2 back into equation 2.1

$$\Delta G_v = \Delta H_v - T\Delta S_v$$

$$= \Delta H_v - \frac{T\Delta H_v}{T_o}$$

$$= \frac{\Delta H_v(T_o - T)}{T_o} = \frac{\Delta H_v \Delta T}{T_o} \quad \text{.......... 2.3}$$

where ΔH_v = Latent of Transformation, γ→α/unit volume of α.

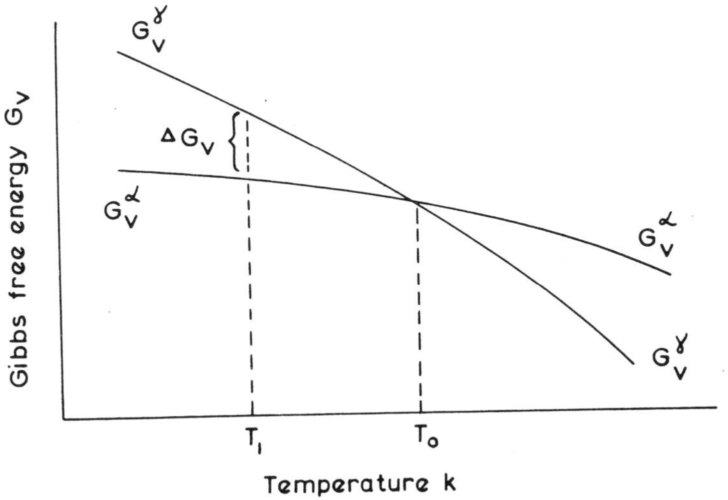

FIG. 2.1 Variation of Gibbs free energy with temperature for two phases α and γ.

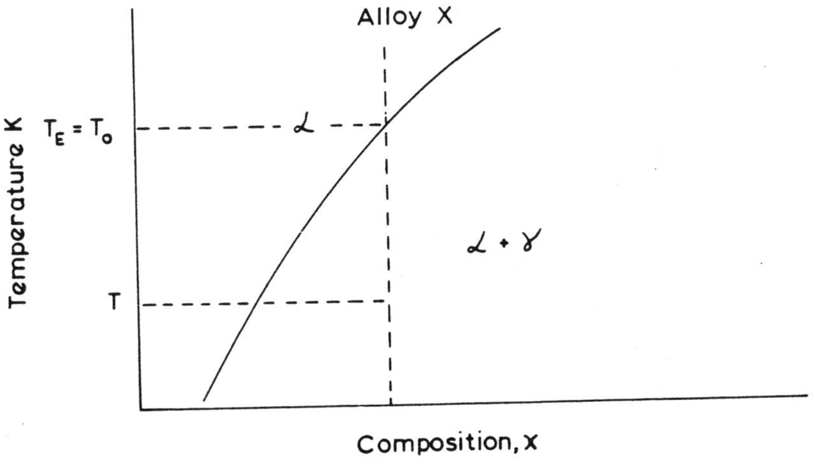

FIG. 2.2 Illustrating T_o for an age hardening alloy X.

Problem 2.1

A good example where equation 2.3 can be applied is the formation of pearlite from austenite in plain carbon steel.

In this case $T_O = T_E = 723^{o}C + 273^{o}C = 996K$

$\Delta H_v = 82$ J/g

Density of ferrite, α, = 7.86 g/ml

Density of Cementite, Fe_3C = 7.4 g/ml

Ferrite solid solubility = 0.02%

Eutectoid composition = 0.8%

Cementite composition ∓ 6.67%c

In 100 grams of material, applying the lever rule

$$\text{Mass of } \alpha = \frac{6.67 - 0.8}{6.67 - 0.02} \times 100 = 88.27 \text{ g}$$

$$= 88.27/7.86 = 11.23 \text{ ml}$$

$$\text{Mass of } Fe_3C = \frac{0.8 - 0.02}{6.67 - 0.02} = 11.73 \text{ g}$$

$$= 11.73/7.4 = 1.585 \text{ ml}$$

$$\therefore \text{Density of pearlite} = \frac{100}{11.23 + 1.585}$$

$$= 7.8 \text{ g/ml}$$

$$\therefore \Delta H_v = 82 \times 7.8 = 396.6 \text{ J/cc}$$

$$= 0.64 \text{ J/ml}$$

$$\Delta G_v = \Delta H_v - T\Delta S_v$$

$$= \frac{\Delta H_v \Delta T}{T_o} = \frac{0.64\ (996 - T)}{996}$$

$$= 0.64 - 6.43 \times 10^{-4} T \text{ J/ml}$$

Note T is in Kelvin.

In the case of a reaction involving changes in composition such as age hardening illustrated in figure 2.2.

$T_o = T_E$ = equilibrium temperature for alloy X, ie T_o can be equated with the solvus temperature for the particular alloy being considered.

Where transformation occurs by a reaction involving no change in composition such as massive transformations or martensitic transformation we can define a metastable equilibrium temperature T_o at which $G^\alpha = G^\gamma$. In this case G^α and G^γ are the free energies of the supersaturated phases α and γ of the same composition. Thus transformation involving no change in composition cannot occur until the supersaturated γ is cooled below T_o. Transformations in Fe-Ni alloys are good examples of this behaviour, figure 2.3 (2.2).

Kaufman and Cohen (2.3) have shown that in this case

$$\Delta G_v \text{ at } M_s = -\Delta G_v \text{ at } A_s \quad \ldots\ldots \quad 2.4$$

$$\therefore \quad T_o = \tfrac{1}{2}(M_s + A_s) \quad \ldots\ldots \quad 2.5$$

In the absence of transformation temperatures, M_s and A_s for the forward and reverse transformations, T_o can be found from the equilibrium phase diagram and lies approximately half way between the equilibrium solvus lines at a particular temperature.

However at large degrees of undercooling ΔG_v has to be computed from a knowledge of the variation of the specific heat of parent and product with temperature and the latent heat of transformation.

2.2 Computation of ΔG_v from Specific Heat and Latent Heat Data

For a phase i and specific heat C_p^i, then at temperature T

$$\text{Enthalpy } H_T^i = H_{298}^i + \int_{298}^{T} C_p^i \, dT \quad \ldots\ldots \quad 2.6$$

$$\text{and Entropy } S_T^i = S_{298}^i + \int_{298}^{T} \frac{C_p^i}{T} \, dT \quad \ldots\ldots \quad 2.7$$

Hence for the $\gamma \to \alpha$ transformation with an intermediate phase β as in figure 2.4 enthalpy change $\Delta H^{\gamma\to\alpha}$ at temperature T below T_c, is given by:-

$$\begin{aligned}
\Delta H^{\gamma\to\alpha} &= H^{(\text{PRODUCTS})} - H^{(\text{REACTANTS})} \\
&= H^\alpha - H^\gamma \\
&= +H^\beta - H^\gamma + H^\alpha - H^\beta \\
&= H_{298}^\beta + \int_{298}^{T} C_p^\beta \, dT - H_{298}^\gamma - \int_{298}^{T} C_p^\gamma \, dT \\
&\quad + H_{298}^\alpha + \int_{298}^{\alpha} C_p^\alpha \, dT - H_{298}^\beta - \int_{298}^{T} C_p^\beta \, dT
\end{aligned}$$

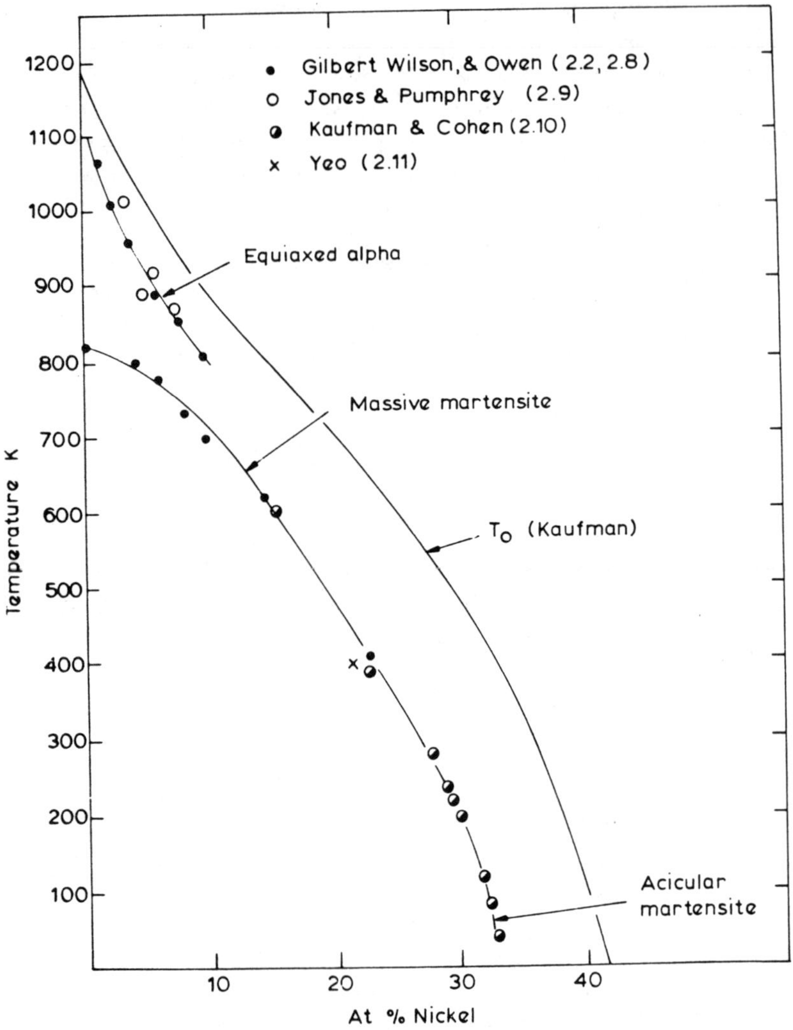

FIG. 2.3 Transformation start temperature on continuosly cooling iron - nickel alloys.

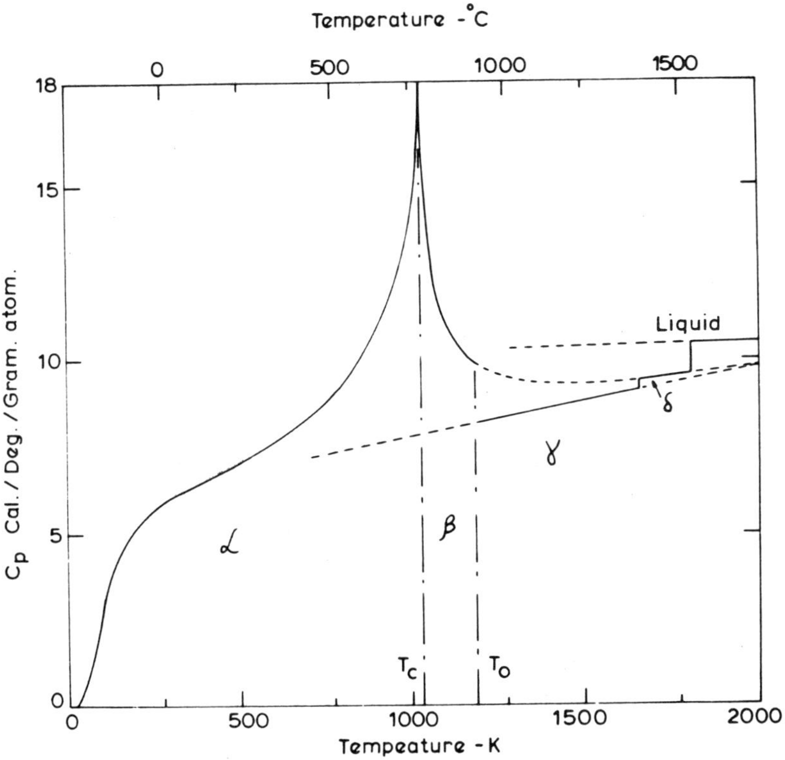

FIG. 2.4 Selected heat capacities of iron (2.4)

$$= \Delta H_{298}^{\gamma\to\beta} + \int_{298}^{T} \Delta C_p^{\gamma\to\beta}\, dT + \Delta H^{\beta\to\alpha} + \int_{298}^{T} \Delta C_p^{\beta\to\alpha}\, dT \quad \ldots\ldots \quad 2.8$$

where $\Delta C_p^{\gamma\to\beta} = C_p^{\beta} - C_p^{\gamma}$

$$\Delta C_p^{\beta\to\alpha} = C_p^{\alpha} - C_p^{\beta}$$

Similarly entropy change for $\gamma\to\alpha$ transformation

$$\Delta S^{\gamma\to\alpha} = \Delta S_{298}^{\gamma\to\beta} + \int_{298}^{T} \frac{\Delta C_p^{\gamma\to\beta}}{T}\, dT + \Delta S_{298}^{\beta\to\alpha}$$

$$+ \int_{298}^{T} \frac{\Delta C_p^{\beta\to\alpha}}{T}\, dT \quad \ldots\ldots \quad 2.9$$

Now at T_o,

$$\Delta H_{T_o}^{\gamma\to\beta} = \text{Heat of transformation, } \gamma\to\beta$$

$$= H_{T_o}^{\beta} - H_{T_o}^{\gamma}$$

$$= H_{298}^{\beta} - H_{298}^{\gamma} + \int_{298}^{T_o} C_p^{\beta}\, dT - \int_{298}^{T_o} C_p^{\gamma}\, dT$$

$$= \Delta H_{298}^{\gamma\to\beta} + \int_{298}^{T_o} \Delta C_p^{\gamma\to\beta}\, dT$$

$$= \Delta H_{298}^{\gamma\to\beta} + \int_{298}^{T} \Delta C_p^{\gamma\to\beta}\, dT + \int_{T}^{T_o} \Delta C_p^{\gamma\to\beta}\, dT$$

∴ Re-arranging

$$\Delta H_{298}^{\gamma\to\beta} + \int_{298}^{T} \Delta C_p^{\gamma\to\beta}\, dT = \Delta H_{T_o}^{\gamma\to\beta} - \int_{T}^{T_o} \Delta C_p^{\gamma\to\beta}\, dT \quad \ldots\ldots \quad 2.10$$

Similarly $\Delta S_{T_o}^{\gamma\to\beta}$ = Entropy of transformation, $\gamma\to\beta$

$$= \Delta S_{298}^{\gamma\to\beta} + \int_{298}^{T_o} \frac{\Delta C_p^{\gamma\to\beta}}{T_o}\, dT$$

$$= \Delta S_{298}^{\gamma\to\beta} + \int_{298}^{T} \frac{\Delta C_p^{\gamma\to\beta}}{T}\, dT + \int_{T}^{T_o} \frac{\Delta C_p^{\gamma\to\beta}}{T}\, dT$$

$$\therefore \Delta S_{298}^{\gamma\to\beta} + \int_{298}^{T} \frac{\Delta C_p^{\gamma\to\beta}}{T}\, dT = \Delta S_{T_o}^{\gamma\to\beta} - \int_{T}^{T_o} \frac{\Delta C_p^{\gamma\to\beta}}{T}\, dT \quad \ldots\ldots \quad 2.11$$

Also since $\Delta G^{\gamma\to\beta} = \Delta H^{\gamma\to\beta} - T\Delta S^{\gamma\to\beta} = 0$ at T_o

$$\therefore \Delta S_{T_o}^{\gamma\to\beta} = \frac{\Delta H_{T_o}^{\gamma\to\beta}}{T_o} \quad \ldots\ldots \quad 2.12$$

$\therefore$ Substituting equation 2.11 in 2.12

$$\Delta S_{298}^{\gamma\to\beta} + \int_{298}^{T} \frac{\Delta C_p^{\gamma\to\beta}}{T} dT = \frac{\Delta H_{T_o}^{\gamma\to\beta}}{T_o} - \int_{T}^{T_o} \frac{\Delta C_p^{\gamma\to\beta}}{T} dT \quad \text{............ 2.13}$$

Similarly at T_c

$$\Delta H_{298}^{\beta\to\alpha} + \int_{298}^{T} \Delta C_p^{\beta\to\alpha} dT = \Delta H_{T_c}^{\beta\to\alpha} - \int_{T}^{T_c} \Delta C_p^{\beta\to\alpha} dT \quad \text{.............. 2.14}$$

and $$\Delta S_{298}^{\beta\to\alpha} + \int_{298}^{T} \frac{\Delta C_p^{\beta\to\alpha}}{T} dT = \frac{\Delta H_{T_c}^{\beta\to\alpha}}{T_c} - \int_{T}^{T_c} \frac{\Delta C_p^{\beta\to\alpha}}{T} dT \quad \text{............. 2.15}$$

where $\Delta H_{T_c}^{\beta\to\alpha}$ = Heat of Transformation, $\beta\to\alpha$.

Hence the chemical driving force $\Delta G^{\gamma\to\alpha}$ for $\gamma\to\alpha$ transformation at temperature T is given by

$$\Delta G^{\gamma\to\alpha} = \Delta H^{\gamma\to\alpha} - T\Delta S^{\gamma\to\alpha}$$

$$= \Delta H^{\gamma\to\beta} + \Delta H^{\beta\to\alpha} - T\Delta S^{\gamma\to\beta} - T\Delta S^{\beta\to\alpha} \quad \text{.................. 2.16}$$

Substituting equations (2.8) and (2.9) into (2.16)

$$\Delta G^{\gamma\to\alpha} = \Delta H_{298}^{\gamma\to\beta} + \int_{298}^{T} \Delta C_p^{\gamma\to\beta} dT + \Delta H_{298}^{\beta\to\alpha} + \int_{298}^{T} \Delta C_p^{\beta\to\alpha} dT$$

$$- T\Delta S_{298}^{\gamma\to\beta} - T\int_{298}^{T} \frac{\Delta C_p^{\gamma\to\beta}}{T} dT - T\Delta S_{298}^{\beta\to\alpha} - T\int_{298}^{T} \frac{\Delta C_p^{\beta\to\alpha}}{T} dT$$

Which simplifies to, using equations (2.10), (2.13), (2.14) and (2.15):-

$$\Delta G^{\gamma\to\alpha} = \Delta H_{T_o}^{\gamma\to\beta} - \int_{T}^{T_o} \Delta C_p^{\gamma\to\beta} dT - T\frac{\Delta H_{T_o}^{\gamma\to\beta}}{T_o} + T\int_{T}^{T_o} \frac{\Delta C_p^{\gamma\to\beta}}{T} dT$$

$$+ \Delta H_{T_c}^{\beta\to\alpha} - \int_{T}^{T_o} \Delta C_p^{\beta\to\alpha} dT - T\frac{\Delta H_{T_c}^{\beta\to\alpha}}{T_c} + T\int_{T}^{T_c} \frac{\Delta C_p^{\beta\to\alpha}}{T} dT \text{... 2.17}$$

For the simple transformation $\gamma\to\beta$ equation 2.17 reduces to

$$\Delta G^{\gamma\to\beta} = \Delta H_{T_o}^{\gamma\to\beta} - \int_{T}^{T_o} \Delta C_p^{\gamma\to\beta} dT - T\frac{\Delta H_{T_o}^{\gamma\to\beta}}{T_o} + T\int_{T}^{T_o} \frac{\Delta C_p^{\gamma\to\beta}}{T} dT \text{...... 2.18}$$

Problem 2.2

Calculate the chemical driving force for the gamma to alpha transformation in pure iron at 690°C given the following data (2.5):

(i) For α, β, δ iron (25-1536°C)

$$C_p = 8.873 + 1.474 \times 10^{-3}T - 56.92T^{-\frac{1}{2}} \text{ cals/deg g-atom}$$

(ii) For γ iron (914-1391°C)

$$C_p = 5.85 + 2.02 \times 10^{-3}T \text{ cal/deg g-atom}$$

(iii) Transformation α→β (760°C)

$$\Delta H = 1216 \text{ cals/g-atom}$$

(iv) Transformation β→γ (914°C)

$$\Delta H = 156 \text{ cals/g-atom}$$

In equation 2.17 $\Delta H_{T_o}^{\gamma\to\beta} = -156$ cals/g-atom

$$T_o = 914^\circ C + 273 = 1187 \text{ K} \qquad T = 690 + 273 = 963 \text{ K}$$

$$\Delta C_p^{\gamma\to\beta} = C_p^{\beta} - C_p^{\gamma}$$

$$= 8.873 + 1.474 \times 10^{-3}T - 56.92T^{-\frac{1}{2}} - 5.85 - 2.02 \times 10^{-3}T$$

$$= 3.023 - 5.46 \times 10^{-4}T - 56.92T^{-\frac{1}{2}}$$

$$\therefore \int_T^{T_o} \Delta C_p^{\gamma\to\beta}\, dT = \int_{963}^{1187} 3.023 - 5.46 \times 10^{-4}T - 56.92T^{-\frac{1}{2}}$$

$$= \left[3.023T - 5.46 \times 10^{-4} \times \tfrac{1}{2}T^2 - 56.92 \times 2T^{\frac{1}{2}}\right]_{963}^{1187}$$

$$= \left[-7.183 \times 10^2 + 8.747 \times 10^2\right]$$

$$= \underline{1.562 \times 10^2 \text{ cals/g-atom}}$$

$$T\frac{\Delta H_{T_o}^{\gamma\to\beta}}{T_o} = \frac{-963 \times 156}{1187} = \underline{-126.6 \text{ cals/g-atom}}$$

$$T\int_T^{T_o} \frac{\Delta C_p^{\gamma\to\beta}}{T}\, dT = 963\int_{963}^{1187} \left(\frac{3.023}{T} - 5.46 \times 10^{-4} - 56.92T^{-3/2}\right) dT$$

$$= 963\left[3.023\ln T - 5.46 \times 10^{-4}T + 113.84T^{-\frac{1}{2}}\right]_{963}^{1187}$$

$$= 963\left[24.06 - 23.91\right]$$

$$= \underline{143.66 \text{ cals/gm-atom}}$$

$\Delta H_{T_c}^{\beta\to\alpha}$ = $\underline{-1216 \text{ cals/gm-atom}}$

$\Delta C_p^{\beta\to\alpha} = 0$ $\qquad\qquad$ $T_c = 760 + 273 = 1033$

$$T \frac{\Delta H_{T_c}^{\beta\to\alpha}}{T_c} = -963 \times \frac{1216}{1033} \qquad = \underline{-1133.6 \text{ cals/gm-atom}}$$

$\therefore$ $\Delta G^{\gamma\to\alpha}$ at 690°C

$= -156 - 156.2 + 126.6 + 143.7 - 1216 + 1133.6$

$= \underline{\underline{-124.3 \text{ cals/gm-atom}}}$

$= \dfrac{-124.3}{55.83}$ cals/gm

$= \dfrac{-124.3}{55.85} \times 7.86$ cals/ml $\qquad$ Density of Fe = 7.86 g/ml

$= \dfrac{-124.3}{55.85} \times 7.86 \times 4.1868$ J/ml

$= \underline{\underline{-73.2 \text{ J/ml}}}$

The value of 124.3 cals/g-atom for $\Delta G^{\gamma\to\alpha}$ at 690°C may be compared with the value of 108.3 cals/gm atom given by the data of Fisher (2.6) and the value of 102.6 cals/gm-atom given by the data of Kaufman et al (2.7).

References

2.1 C.Zener, Trans AIME, 1946, Vol 167, 950.

2.2 W.S. Owen, E.A. Wilson and T. Bell, "High Strength Materials", Editor V.F. Zackay, J. Wiley, New York 1965, p.167.

2.3 C. Kaufman and M. Cohen, "Thermodynamics and Kinetics of Martensitic Transformations", Progress in Metal Physics, 1958, Vol.7, p.165-246.

2.4 L.S. Darken and R.P. Smith, J. Ind. and Eng. Chem., 1951, Vol.43, p.1815-20.

2.5 W.A. Dench and O. Kubaschewski, JISI, 1963, pp.140-142.

2.6 J.C. Fisher, Metals Transactions 1949, Vol.185, p.688.

2.7 L. Kaufman, E.V. Clougherty and R.J. Weiss, Acta Met., 1963, Vol.11, p.323-335.

2.8 A. Gilbert and W.S. Owen, 1962, Acta Met., Vol.10, p.45.

2.9 F.W. Jones and W.I. Pumphrey, JISI, 1949, Vol.163, p.121.

2.10 L. Kaufman and M. Cohen, Trans AIME, 1956, Vol.206, p.1393.

2.11 R.G. Yeo, Trans AIME 1962, Vol 224, p.1222.
Trans. ASM, 1964, Vol.57, p.98.

3. The Arrhenius Relationship

3.1 Metastability

As a mechanical analogue of thermodynamic phase metastability consider the potential energy of a block of mass m resting on a horizontal plane, figure 3.1.

The potential energy, mgh, of the block varies with the angle of tilt θ as shown in figure 3.2.

Position (1) is a metastable equilibrium position since although the block does not have the minimum possible potential energy, any small change in θ results in an increase in the potential energy of the block which opposes the change in θ.

Position (3) is the stable equilibrium position of the block, since it possesses the minimum possible potential energy with respect to the horizontal surface.

Position (2) is a position of unstable equilibrium since any slight change in θ produces a decrease in potential energy and the block tends to either position (1) or position (3).

For the block to move from position (1) to position (3) energy $Q = mg\,\Delta h$ has to be supplied to the block.

Note energy $H = mg\,(h_1 - h_2)$ is released when the block reaches position (3).

Similar situations in the case of movements of atoms from one state to another and are illustrated in figure 3.3.

Thus for example an atom may be diffusing from a lattice site (i) into a vacant lattice site (iii). Alternatively an atom may be moving from one metastable phase γ, (i), to join the equilibrium phase α, (iii), during transformation of γ to α.

Energy Q has to be supplied to the atom before it can move from state (i) to state (ii) and latent heat of transformation H is released on reaching state (iii).

Position (ii) is known as the transition state or activated state and corresponds with position (2) in the mechanical analogue.

3.2 Rates of Reaction

The number of atoms moving/unit time from state (i) to state (iii) depends

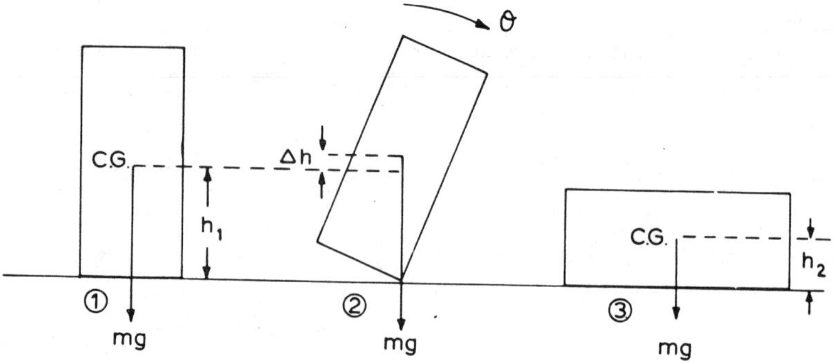

FIG. 3.1 Stability of a block on a surface.
Mechanical analogue of thermodynamic phase stability.

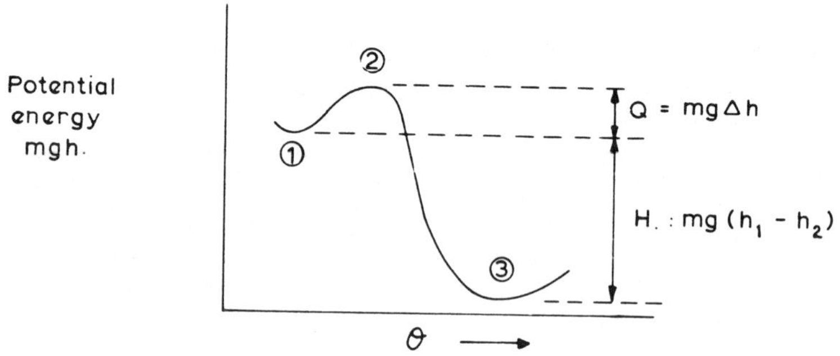

FIG. 3.2 Variation of potential energy of block in figure 3.1 with angle of tilt θ

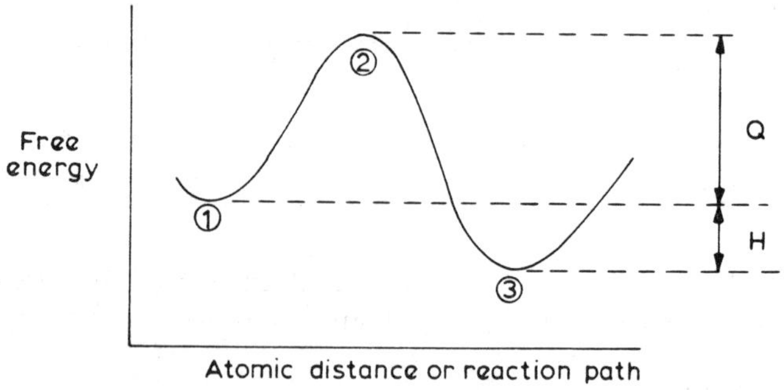

FIG. 3.3 Variation of free energy with thermodynamic state

upon

(a) the frequency ν which atoms attempt to scale the barrier,

(b) the number of atoms n acquiring sufficient energy Q to surmount the activation energy barrier. This is given by Boltzmann statistics as

$$N \exp - \frac{Q}{kT}$$

where k = Boltzmann constant

T = absolute temperature in kelvin $^{o}C + 273$

N = No of atoms in state (i).

(c) the probability p, that the atom is moving in the right direction to scale the barrier.

Thus for one dimensional movement, figure 3.4, $p = \frac{1}{2}$, since for half the time the atom is moving in the opposite direction to the reaction path.

For three dimensional movement, figure 3.5, $p \sim 1/6$.

Therefore if y=fraction of atoms moving from state (i) to state (iii) $= \frac{n}{N}$ then, Rate of raction $= \frac{dy}{dt} = p\nu \exp - \frac{Q}{RT}$ 3.1

Now p is related to the change in entropy ΔS on moving from position (i) to the transition state (ii) and is given by

$$p = \exp \frac{\Delta S}{k} \quad \text{..........} \quad 3.2$$

$$\therefore \frac{dy}{dt} = \nu \exp \frac{\Delta S}{k} \exp - \frac{Q}{kT} \quad \text{..........} \quad 3.3$$

$$= A \exp - \frac{Q}{kT} \quad \text{..........} \quad 3.4$$

A is known as the frequency factor and Q the activation energy for the transition from state (i) to state (iii) while equation 3.4 is known as the Arrhenius equation.

In logarithmic form, equation becomes

$$\ln \left(\frac{dy}{dt}\right) = \ln A - \frac{Q}{k} \cdot \frac{1}{T} \quad \text{..........} \quad 3.5$$

Thus if Q and A are independent of temperature then a plot of $\ln \left(\frac{dy}{dt}\right)$ versus 1/T is linear with slope - Q/k and intercept ln A.

Frequently one finds in metallurgical reactions (see section 4.2.4):-

$$\frac{dy}{dt} \alpha \frac{1}{t_y} \quad \text{..........} \quad 3.6$$

where t_y = time to form a given fraction y of transformation product at temperature T.

Hence $\frac{1}{t_y} = B \exp - \frac{Q}{kT}$ 3.7

and $\ln 1 - \ln t_y = B - \frac{Q}{k} \cdot \frac{1}{T}$

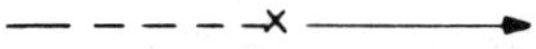

FIG. 3.4 Illustrating one dimensional movement in attempting to scale activation energy barrier.

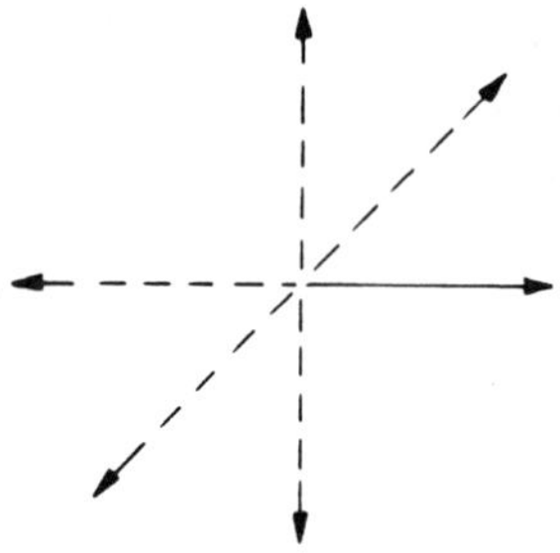

FIG. 3.5 Illustrating 3 dimensional movement in attempting to scale activation energy barrier.

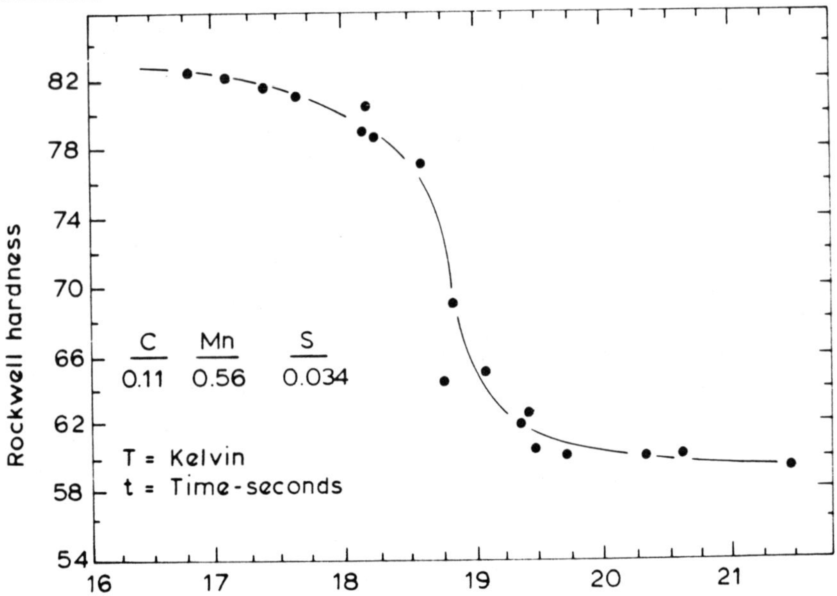

FIG. 3.6 Annealing curve for mild steel.

$$\text{i.e. } \ln t_y = \frac{Q}{k} \cdot \frac{1}{T} - \ln B \quad \ldots\ldots \quad 3.8$$

where B incorporates the frequency factor A and the constant of proportionality in equation 3.7.

Equation 3.8 is often used in determining the activation energy for phase transformations.

3.3 An alternative treatment

If the number of atoms in state (ii) is n* and the number of atoms in state (i) is N, therefore the reaction:

State (i) → State (ii)

(N) → (n*)

The rate constant $K = \frac{n^*}{N}$

$$\text{and } \Delta G^* = -\,kT \ln K \quad \ldots\ldots \quad 3.9$$

where ΔG* = difference in Gibbs free energy/atom between states (i) and states (ii).

$$\therefore \frac{n^*}{N} = \exp - \frac{\Delta G^*}{kT} \quad \ldots\ldots \quad 3.10$$

$$\therefore \text{Rate of reaction } \frac{dy}{dt} \,\alpha\, n^* = N \exp - \frac{\Delta G^*}{kT} \quad \ldots\ldots \quad 3.11$$

Now $\Delta G^* = \Delta H^* - T\Delta S^*$

$$\therefore \frac{dy}{dt} \,\alpha\, \exp - \frac{\Delta S^*}{k} \exp - \frac{\Delta H^*}{kT} \quad \ldots\ldots \quad 3.12$$

ΔH* can be identified with Q and hence assuming ΔH* and ΔS* do not vary with temperature.

$$\frac{dy}{dt} = A \exp - \frac{Q}{kT} \quad \ldots\ldots \quad 3.13 \; (3.4)$$

3.4 Units of the activation energy Q

If Q is in J/atom then we use

$$\frac{dy}{dt} = A \exp - \frac{Q}{kT}$$

where k = Boltzmann constant

$= 1.38044 \text{ J atom}^{-1} \text{ K}^{-1}$

If Q is in J/mol then we use

$$\frac{dy}{dt} = A \exp - \frac{Q}{RT}$$

where R = Gas constant

$= 8.314 \text{ J mol}^{-1}\text{K}^{-1}$

Problem 3.1 If after roller levelling mild steel, the yield point starts to return after 6 months, at 22°C, calculate how long it takes the yield point to return at 127°C.

Activation energy for diffusion of N in α iron = 78 kJ/mol.

We use the time dependent form of the Arrhenius equation, 3.7:

$$\frac{1}{t_y} = B \exp - \frac{Q}{RT}$$

Thus if the yield point returns after times t_1 and t_2 at temperatures T_1 and T_2 kelvin respectively

$$\frac{1}{t_1} = B \exp - \frac{Q}{RT_1} \qquad \text{and} \qquad \frac{1}{t_2} = B \exp - \frac{Q}{RT_2}$$

$$\therefore \frac{t_2}{t_1} = \frac{\exp - Q/RT_1}{\exp - Q/RT_2} = \exp - \frac{Q}{R}\{\frac{1}{T_1} - \frac{1}{T_2}\}$$

or $$\ln \frac{t_2}{t_1} = - \frac{Q}{R}\{\frac{1}{T_1} - \frac{1}{T_2}\} = \frac{Q}{R}\{\frac{1}{T_2} - \frac{1}{T_1}\}$$

Now t_1 = 6 months = 26 x 7 x 24 x 60 mins

T_1 = 22°C = 295 K

t_2 = ? T_2 = 27°C = 300 K

$$\therefore \ln \frac{t_1}{t_2} = \frac{78{,}000}{8.314}\{\frac{1}{300} - \frac{1}{295}\} = -0.53$$

$$\frac{t_2}{t_1} = 0.5886$$

t_2 = 0.5886 x 26 x 7 x 24 x 60 mins

= <u>15.4 mins</u>

<u>Problem 3.2</u> Q.1 Associateship in Metallurgy, Physical Metallurgy B, June 1977, Sheffield City Polytechnic.

Figure 3.6 shows an annealing curve for mild steel sheet. To which physical processes do the various parts of this curve correspond?

Discuss the commercial feasibility of producing annealed mild steel sheet with a hardness value of 70 Rockwell.

From the annealing curve calculate the activation energy required for approximately 50% recrystallisation.

R = 8.314 J/mol.K. $\log_e x = 2.303 \log_{10} x$

At high hardness levels ∿ 82-78 Rockwell, little change in mechanical properties is occurring and this corresponds to a recovery process.

While over the next part of the annealing curve, 78-62 Rockwell a rapid decrease in hardness occurs and recrystallisation is taking place.

At longer times and higher temperatures, little change in hardness occurs and grain growth is occurring.

Because of the steep fall in hardness during recrystallisation at a particular value of $T(20 + \log_{10}t)$ it would be very difficult to control the annealing process to give a hardness of 70 Rockwell, ie slight variations in time and temperature of annealing would easily result in a hardness of less than 70 Rockwell.

At 70 Rockwell this corresponds to 50% recrystallisation and $T(20 + \log t) \times 10^{-3} = 18.75$.

$$\text{ie} \quad 20 + \log t = \frac{18.75 \times 10^3}{T} = \frac{155.89 \times 10^3}{RT}$$

$$\log 10^{20} t = \frac{156 \times 10^3}{RT}$$

$$\ln 10^{20} t = \frac{156 \times 10^3}{RT} \times 2.303 = \frac{359 \times 10^3}{RT}$$

$$\text{ie} \quad \frac{1}{t} = 10^{20} \exp - \frac{359 \times 10^3}{RT}$$

$$\text{cf} \quad \frac{1}{t} = A \exp - \frac{Q}{RT}$$

∴ Activation energy for 50% recrystallisation Q = 359 kJ/mol

and A = 10^{20} seconds^{-1}

Problem 3.3 Determine graphically, or otherwise, the recrystallisation temperature for cold worked copper from the following data. Assume an annealing time of one hour.

Data

Time for 100% recrystallisation at 300°C = 50 h

Time for 100% recrystallisation at 350°C = 6 h

Problem 3.4 A steel develops a case depth of 1.25 mm after carburising for 6 hours at 925°C. Assuming that the relationship between case depth x, carburising time t, and diffusion coefficient D is given by $x^2 = 4\,Dt$ determine the case depth after carburising for 6 hours at 1010°C.

Activation energy for diffusion of carbon in austenite = 153 kJ/mol.

Problem 3.5 The following data were obtained for the time of commencement of the austenite to bainite transformation at various temperatures in a 0.8% carbon steel.

Temperature °C	265	290	308	330	360	375	410
Time (seconds)	840	670	600	420	220	112	35

Calculate the activation energies for the formation of upper and lower bainite. To what physical processes should these energies correspond?

4. The Isothermal Rate Equation

During isothermal transformation the volume fraction of second phase y increases with time t. In practice it is often found that y varies with t in a sigmoid fashion. This chapter describes the equation governing this form of nucleation and growth and the important kinetic properties of this equation.

4.1 Johnson and Mehl Equation

The following derivation is due to Johnson and Mehl (41) and describes the nucleation and growth of random nucleated spherical "nodules".

Johnson and Mehl made the following assumptions in their derivation:

1. Nucleation and growth occur simultaneously throughout the period of transformation.
2. Nucleation is completely random throughout the transforming material. That is nucleation occurs homogeneously rather than at preferred sites, (i.e. heterogeneous nucleation).
3. The nucleation rate, $I = \frac{dn}{dt}$ = number of nuclei/unit volume/unit time, is constant throughout the process.
4. There is a constant linear growth rate, $G = \frac{dx}{dt}$ of the second phase particles during transformation.
5. Particles of second phase nucleate and grow as spheres.

Consider a spherical nodule nucleated at time τ.

At time t, time of growth = $(t - \tau)$

$\therefore$ Radius r of nodule at time t = $G(t - \tau)$

and hence volume of nodule at time t $= \frac{4}{3}\pi r^3$

$$= \frac{4}{3}\pi G^3(t - \tau)^3$$

No. of nodules nucleated in time $d\tau$ in volume of material V = $VId\tau$.

$\therefore$ Volume of second phase formed in time $d\tau = VI\frac{4}{3}\pi G^3(t - \tau)^3 d\tau$

Hence total volume of new phase grown in time t = $v_x = \int_{\tau=0}^{\tau=t} VI\frac{4}{3}\pi G^3(t-\tau)^3 d\tau$

and Volume Fraction $y_x = \frac{v_x}{V} = \int_{\tau=0}^{\tau=t} I\frac{4}{3}\pi G^3(t-\tau)^3 d\tau$

ie $dy_x = \frac{4}{3}\pi IG^3(t - \tau)^3 d\tau$ 4.1

However this equation does not take into account two factors:-

1. Nucleation is no longer occuring in the volume of material already transformed. We have assumed the nucleation rate is constant throughout the volume of material V. In fact we should have used the volume

(V - v).

2. Impingement of growing nodules, for example in figure 4.1 the shaded area has been counted twice.

For this reason y_x has the suffix x and is known as the extended volume fraction compared with the real volume fraction y.

There is a relationships between y_x and y. The clearest treatment of this relationship known to the author is due to Christian (4.2) and is as follows:

Let v = real volume of material transformed

v_x = extended volume transformed

V = volume of material or specimen.

"Consider a small random region of which a fraction {1 - v/V} remains untransformed at time t. During a further time dt, the extended volume of second phase in the region will increase by dv_x and the true volume by dv. Of the new elements which make up dv_x, a fraction {1 - v/V}, ie (1 - y) on the average will lie in previously untransformed material and thus contribute to dv, whilst the remainder of dv_x will be in already transformed material."

$$\therefore\ dv = \{1 - \frac{v}{V}\}dv_x$$

$$= (1 - y)dv_x$$

$$\text{ie } dy = (1 - y)dy_x \quad \text{.......................} \quad 4.2$$

Note these arguments consider that nucleation is occurring randomly throughout the matrix.

Therefore returning to equation 4.1

$$dy_x = 4/3\pi IG^3(t - \tau)^3 d\tau$$

$$\therefore\ dy = (1 - y)dy_x$$

$$= (1 - y)\{4/3\pi IG^3(t - \tau)^3 d\tau\}$$

$$\text{ie } \frac{dy}{(1-y)} = 4/3\pi IG^3(t - \tau)^3 d\tau$$

Integrating

$$\int_{y=0}^{y} \frac{dy}{(1-y)} = \int_{\tau=0}^{\tau=t} 4/3\pi IG^3(t - \tau)^3 d\tau$$

$$\left[-\ln(1 - y)\right]_{y=0}^{y} = \left[-1/3\pi IG^3(t - \tau)^4\right]_{\tau=0}^{\tau=t}$$

$$\ln(1 - y) = -1/3\pi IG^3 t^4$$

$$\text{ie} \quad y = 1 - \exp - \{1/3\pi IG^3 t^4\} \quad \text{.......................} \quad 4.3$$

Thus we may expect a general relationship of the form:

$$y = 1 - \exp - (kt)^n \quad \ldots \quad 4.4$$

to apply to most transformations.

Problem 4.1 Q.3 BSc (Hons) Final Year Physical Metallurgy Paper, June 1978, Sheffield City Polytechnic.

State the assumptions made by Johnson and Mehl in their derivation of the isothermal rate equation for the transformation of austenite to pearlite.

Critically discuss these assumptions with respect to the known kinetics and morphology of the pearlite transformation.

Making the same assumptions as Johnson and Mehl, what would be the value of the time exponent "n" in the isothermal rate equation for the formation of a second phase in the form of needles with linear growth predominantly in the long direction?

The assumptions made by Johnson and Mehl are given at the commencement of Section 4.1. Pearlite grows as spherical nodules or to be more accurate semi-spherical nodules nucleated on austenite grain boundaries. So the assumption of a spherical shape for the "nodules" is very good. Also experimental observations show that pearlite nodules grow linearly with time, in agreement with Johnson and Mehl's assumptions. However, pearlite usually nucleates on the austenite grain boundaries at an early stage in the reaction and then proceeds to grow into the austenite grain until it is consumed. Little nucleation of pearlite occurs within the austenite grains. So the assumption of random nucleation and constant nucleation rate throughout the reaction is clearly in error.

In their original paper (4.1), Johnson and Mehl derived the isothermal rate equation for grain boundary nucleation. They found very good agreement between experimental results and theory for the isothermal formation of pearlite.

For linear growth of a needle of the volume of second phase v_x will be proportional to $(t - \tau)$.

Thus on integration to obtain the true volume fraction y, this will become proportional to $(t - \tau)^2$.

$\therefore$ Time exponent $n = \underline{\underline{2}}$.

4.2 Properties of the General Isothermal Rate Equation

$y = 1 - \exp - (kt)^n$ 4.4

y = fraction transformed

k = rate constant and depends on isothermal holding temperature T kelvin

$= A \exp - \frac{Q}{RT}$ where Q = activation energy of transformation, usually that for growth of the transformation product for temperatures below the nose of the 'C' curve on the TTT diagram.

n = time exponent, varying with the time dependence of the nucleation rate and of the growth rate and the shape of the second phase particle. Usually for the same type of transformation (eg Pearlite) at different temperatures 'n' has approximately the same value.

1. The sigmoid shape of the isothermal rate curve, figure 4.2, is in good agreement with the way that phases are known to nucleate and grow.

   That is nucleation occurs slowly at commencement of transformation. Nucleation and growth rates reach a maximum and then the rate of transformation slows down near completion due to impingement and exhaustion of nucleation sites.

   In fact the sigmoid rate curve and corresponding rate equation tends to be typical of many systems whether they be biological or socio-economic where there are limits to growth and birth or innovation (≡ nucleation); see figures 4.3 and 4.4.

2. When plotted on a log (time) axis the shape of the rate curves depends only on 'n' and their position along the axis on k, figure 4.2.

   That is curves with the same value of 'n' can be superimposed by movement along the log (time) axis, which means that $\Delta \log t$ in figure 4.2 does not vary with y.

   This can be proved as follows:

   Re-arranging equation 4.4

$$(1 - y) = \exp - (kt)^n$$

$$\therefore \ln(1 - y) = -(kt)^n$$

$$\text{ie } -\ln(1 - y) = \ln(1/1 - y) = (kt)^n$$

$$\therefore \ln\ln(1/1-y) = n\ln t + n\ln k \quad \text{.................................. 4.5}$$

For two temperatures T_1 and T_2 with rate constants k_1 and k_2 and considering the same value of $y = y_o$

$$\ln\ln 1/(1 - y_o) = n\ln t_1 + n\ln k_1$$

$$\ln\ln 1/(1 - y_o) = n\ln t_2 + n\ln k_2$$

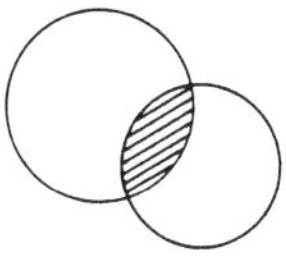

Fig. 4.1 Illustrating impingement of growing spherical modules.

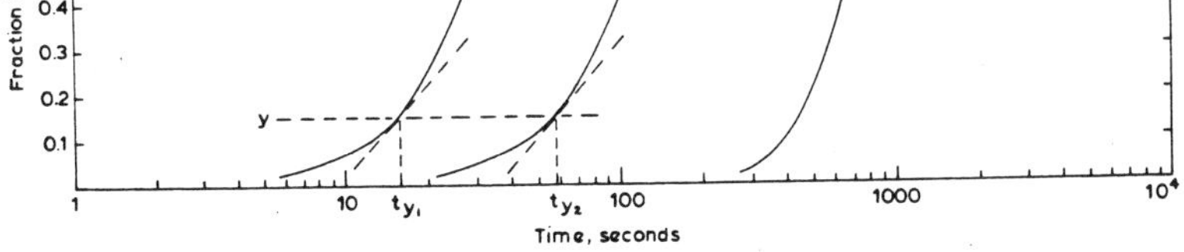

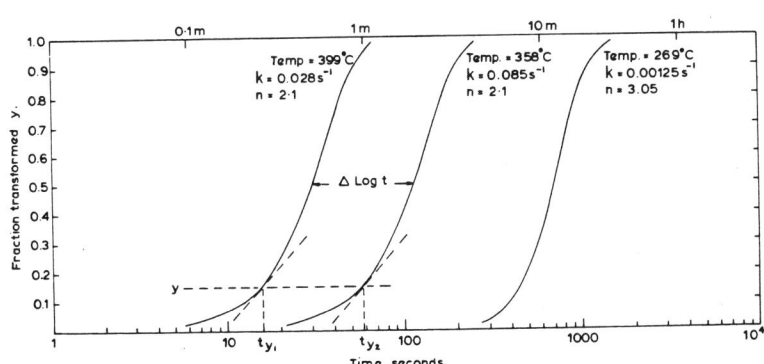

Fig. 4.2 Idealised rate curves, showing how the shape of the curves depends on n and their positions on the log(time) axis on k.

The data is taken from Radcliffe and Rollason (4.4) for isothermal transformation of an Fe 1.07%C alloy. The original data has been altered for transformation greater than 50% to give the same value of n throughout the reaction.

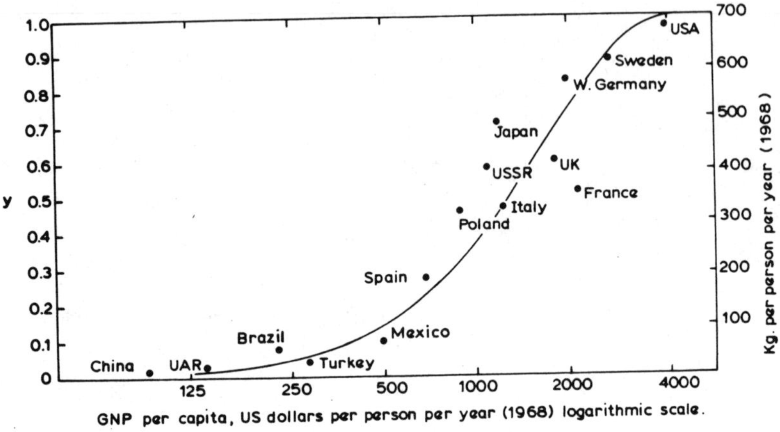

FIG. 4.3 1968 steel consumption per person in various nations of the world. Reference 4.3

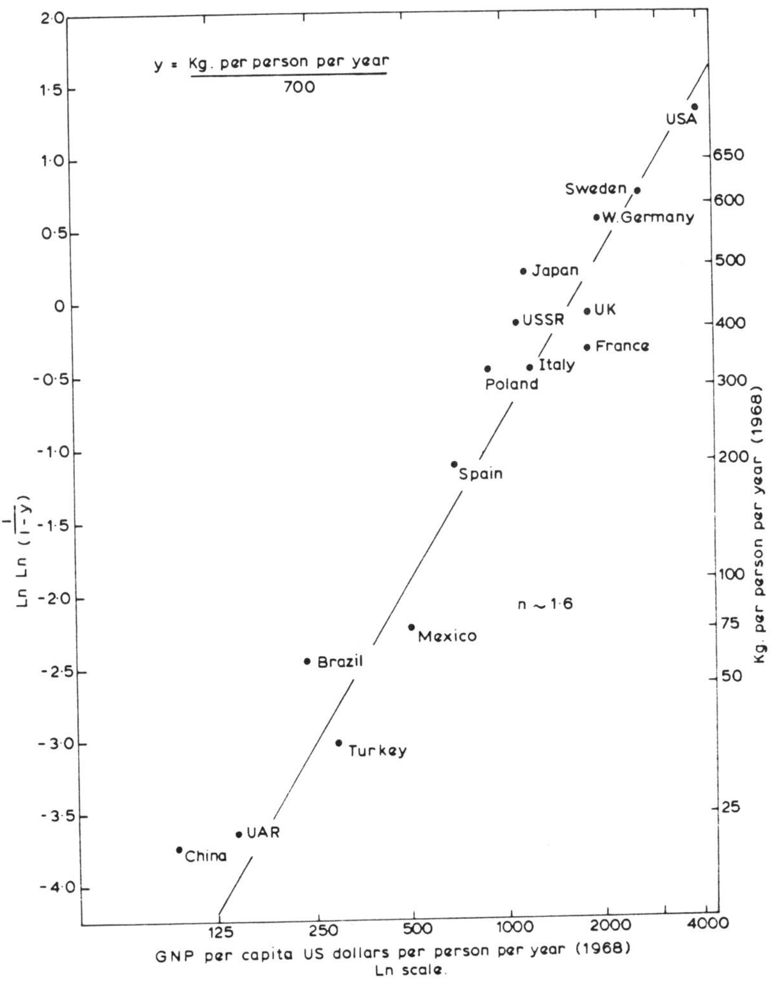

FIG. 4.4 Logarithmic plot of world steel consumption.

Subtracting these two equations

$$n\ln t_1 - n\ln t_2 = n\Delta \ln t = n\ln k_2 - n\ln k_1$$
$$= n\ln\frac{k_2}{k_1}$$

∴ If n is constant at each value of y

$$\Delta \ln t = \ln\frac{k_2}{k_1} \quad \ldots\ldots\ldots\ldots\ldots\ldots\ldots\ldots\ldots\ldots \quad 4.6$$

= constant if k_1 and k_2 are constant.

3. When such rate curves can be superimposed, they are said to be <u>isokinetic</u>. This means that the activation energy for the transformation Q is independent of the fraction transformed y, since:-

Rate constant $k = A \exp - \frac{Q}{RT}$

Taking logarithms, $\ln k = \ln A - \frac{Q}{RT}$

For two temperatures T_1 and T_2 with rate constants k_1 and k_2

$$\ln k_1 = \ln A - \frac{Q}{RT_1}$$

$$\ln k_2 = \ln A - \frac{Q}{RT_2}$$

$$\therefore \ln\frac{k_2}{k_1} = \frac{Q}{R}\left\{\frac{1}{T_1} - \frac{1}{T_2}\right\} = \frac{Q}{R}\Delta\, 1/T$$

From previous equation, 4.6, $\ln\frac{k_2}{k_1} = \Delta \ln t$

$$\therefore Q = R\,\frac{\Delta \ln t}{\Delta\, 1/T}$$

If curves are isokinetic $\Delta \ln t$ is independent of y and therefore Q is independent of y.

4. For isokinetic curves the rate of transformation is inversely proportional to the time taken t_y, to form a particular fraction of transformation product y.

If the isothermal rate curves for a particular reaction are isokinetic, then the slope of the curves when plotted on a ln(time) axis at a particular value of y are equal, see figure 4.2.

ie $\frac{dy}{d\ln t}$ = constant = B at a fixed value of y

But $d\ln t = \frac{dt}{t_y}$ at fixed y

$$\frac{dy}{dt} = \frac{B}{t_y} \propto \frac{1}{t_y}$$

$$= A \exp - \frac{Q}{RT}$$

Hence Q can be obtained from Arrhenius plots using times to a certain fraction, eg 1%, 50% or 100% at different temperatures.

4.3 Determination of k

$y = 1 - \exp - (kt)^n$ 4.4

when $kt = 1$, $y = 1 - \exp -(1)^n = 1 - \exp - (1)$

$= 1 - e^{-1} \qquad = 1 - 1/e$

$= 0.632$

$\therefore \quad k = 1/t$ when $y = 0.632$.

4.4 Determination of n

Taking double logarithms of equation 4.4 gives us equation 4.5.

$\ln\ln (1/1 - y) = n\ln t + n\ln k$ 4.5

cf. $\qquad y \qquad = m.x \ + \ c$

Therefore a plot of $\ln\ln 1/(1 - y)$ versus $\ln t$ is linear and of slope n.

Note either common or natural logarithms can be used provided the same base is used either side of the equation, since:

$\log \ln (1/1 - y) = n\log t + n\log k$ 4.7

and $\log\log (1/1 - y) = n\log t + n\log k - \log 2.303$ 4.8

Problem 4.2 Q.8 Associateship in Metallurgy, Physical Metallurgy 1. June 1974, Sheffield Polytechnic.

The decomposition of austenite in a 0.65% plain carbon steel was followed at 421°C and 358°C by a resistivity technique.

The following table shows the time taken to transform at each temperature to the fraction, y, indicated:

y	0.05	0.10	0.20	0.30	0.40	0.50
t(secs) 421°C	2.4	3.2	4.3	5.4	6.5	7.5
t(secs) 358°C	9.7	12	16	20	25	28

y	0.60	0.70	0.80	0.90
t(secs) 421°C	8.5	9.8	12	16
t(secs) 358°C	33	38	45	56

Show that the above data represent isokinetic reaction and explain the significance of this.

Given that the above data fit the isothermal rate equation $y = 1 - \exp [- (kt)^n]$ determine the value of the rate constant k in this equation for the two temperatures. From these values of k determine the activation energy for the transformation. To what physical process do you think the value of the activation energy corresponds?

Exponential constant e = 2.718

Gas constant R = 8.314 J/mol.K

Log graph paper is provided,

Figure 4.5 shows a plot of the isothermal transformation data on log graph paper. The easiest way of demonstrating the data is isokinetic is to physically measure the distance between the curves with a ruler at various values of y. The results are shown in the table in figure 4.5.

It will be seen Δlogt is approximately constant, 4.75 cms, although there are slight deviations at the beginning and near the end of transformation. This is the usual result obtained in practice.

At y = 1 - 1/e = 0.632

$t_1 = 8.8s$ at $T_1 = 421^oC = 694$ K $\therefore k_1 = 1/8.8 = 0.1136s^{-1}$

$t_2 = 34.5s$ at $T_2 = 358^oC = 631$ K $\therefore k_2 = 1/34.5 = 0.02898s^{-1}$

Now the rate constant k is related to temperature T kelvin by the Arrhenius relationship.

$$k = A \exp - \frac{Q}{RT}$$

ie $$\ln k = \ln A - \frac{Q}{RT}$$

Hence we have for the two temperatures T_1 and T_2 with rate constants k_1 and k_2

$$\ln k_1 = \ln A - \frac{Q}{RT_1}$$

$$\ln k_2 = \ln A - \frac{Q}{RT_2}$$

ie $$\ln k_1 - \ln k_2 = - \frac{Q}{RT_1} + \frac{Q}{RT_2}$$

$$\ln\frac{k1}{k_2} = \frac{Q}{R}\{\frac{1}{T_2} - \frac{1}{T_1}\}$$

$$\therefore \quad Q = \frac{R\ln k_1/k_2}{\{1/T_2 - 1/T_1\}}$$

$$= \frac{8.314 \ln 34.5/8.8}{\{1/631 - 1/694\}}$$

$$= 78.95 \text{ kJ/mol} = \underline{\underline{79 \text{ kJ/mol}}}$$

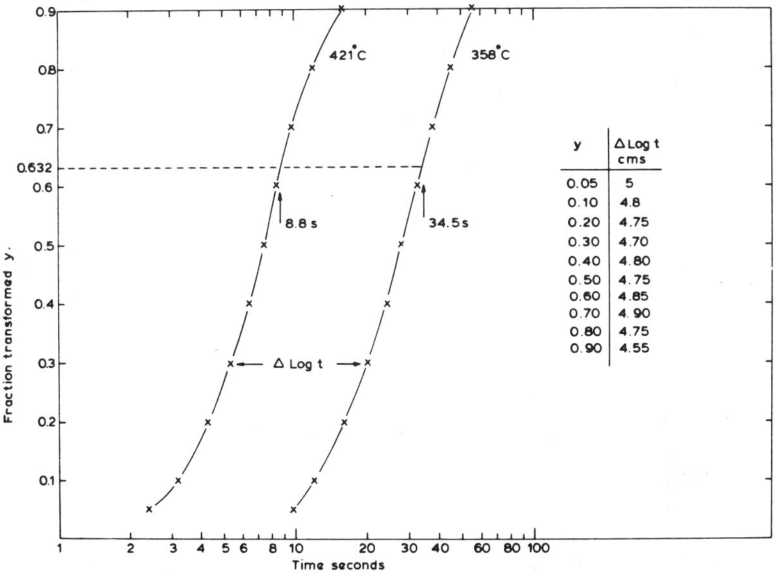

Fig. 4.5 Problem 4.2 Illustrating iso-kinetic behaviour for upper bainite

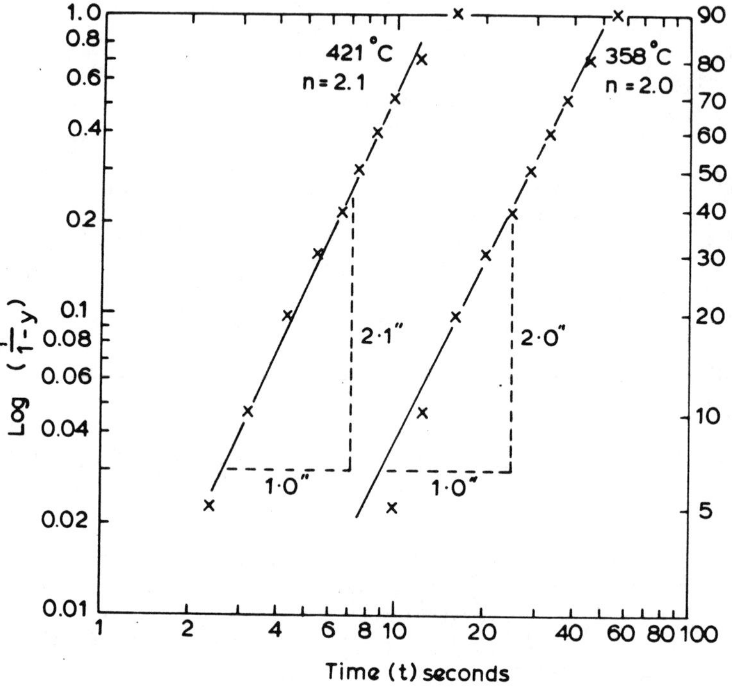

FIG. 4.6 Problem 4.2 Determination of n in isothermal rate equation.

Problem 4.3 Q.4 CNAA BSc (Hons) Final Year, Physical Metallurgy Paper, June 1972, Sheffield Polytechnic

Describe and explain the properties of the isothermal rate equation:

$$y = 1 - \exp-(kt)^n$$

where y = fraction transformed at time t

k = rate constant

n = time exponent

The decomposition of austenite in an Fe–1.07%C alloy at 399°C and 269°C was followed by a resistivity technique. The following values of y and t were recorded:

y		0.05	0.10	0.20	0.30	0.40	0.50	0.60
t(secs)	399°C	10	13	18	21	26	30	35
t(secs)	269°C	340	400	480	570	630	680	730

y		0.70	0.80	0.90
t(secs)	399°C	42	55	70
t(secs)	269°C	820	1000	1300

Determine the value of n for the reactions at 399°C and 269°C. Comment on the values of n that you obtain at these two temperatures

Problem 4.4 Q.3 Associateship in Metallurgy, Physical Metallurgy B, June 1978, Sheffield City Polytechnic.

The decomposition of austenite at 269°C in a plain carbon steel containing 1.07%C was examined by a resistivity technique. The following values of fraction transformed, y, and reaction time, t, were recorded:

y	0.05	0.10	0.20	0.30	0.40	0.50	0.60	0.70
t(in seconds)	340	400	480	570	630	680	760	820

Given that the data follow the isothermal rate equation:

$$y = 1 - \exp-(kt)^n$$

where k = rate constant

n = time exponent

determine the time taken to form 75%, 80%, 85%, 90% and 95% of transformation product.

Explain why the rate of reaction is slowing down near completion.

(Logarithmic graph paper is provided.)

The transformation considered occurs above 350°C, so at these temperatures austenite will be transforming to upper bainite. Classically, it has been suggested that the rate controlling process in the formation of upper bainite is diffusion of carbon in austenite. However, at 0.65%C the activation energy for diffusion of carbon in austenite is ∿138 kJ/mol compared with 79 kJ/mol given by the data in the question. Radcliffe and Rollason conclude that "the formation of bainite is not controlled by a known diffusion process". Nevertheless Radcliffe and Rollason obtained a value of 31.4 kJ/mol for formation of lower bainite in the same 0.65%C steel. This can be compared with the value of 80 kJ/mol for diffusion of carbon in ferrite. Clearly the activation energy for formation of upper bainite is less than that for lower bainite. It could be that the activation energies for formation of upper and lower bainite are less than that for diffusion of carbon in austenite and ferrite, because the diffusion rate is increased by some process.

Determination of n

The table below gives the values of log (1/1 - y) and t:

y	1 - y	log (1/1 - y)	t(358°C)	t(421°C)
0.05	0.95	0.0223	9.7	2.4
0.10	0.90	0.0458	12	3.2
0.20	0.80	0.0969	16	4.3
0.30	0.70	0.155	20	5.4
0.40	0.60	0.222	25	6.5
0.50	0.50	0.301	28	7.5
0.60	0.40	0.400	33	8.5
0.70	0.30	0.522	38	9.8
0.80	0.20	0.700	45	12
0.90	0.10	1.000	56	16

n can be determined from the slope of the graph loglog (1/1-y) versus logt; equation 4.8. By using log/log graph paper unnecessary computation is avoided and moreover the value of n can be obtained by just measuring the slope of the line using a ruler, figure 4.6.

It will be seen that n is approximately the same for the two temperatures. The plot is non-linear near the commencement of transformation and towards the end of transformation. This is a common occurrence with actual experimental data.

References

4.1 W.A. Johnson and R.F. Mehl. Trans AIMME 1939, 135 pp.416-442.

4.2 J.W. Christian. The Theory of Transformation in Metals and Alloys, Pergamon, 1965.

4.3 D.H. Meadows, D.L. Meadows, J Randers, W.W. Behrens III, "Limits to Growth", 1974 Pan Books Ltd, Great Britain p.110.

4.4 S.V. Radcliffe and E.C. Rollason, JISI, 1959, 191 pp.56-65.

5. Homogeneous Nucleation

5.1 Critical nucleus size and activation energy for homogeneous nucleation

We will first consider homogeneous nucleation, that is random nucleation, rather than nucleation at preferred sites or catalysts (ie hetrogeneous nucleation).

Consider the formation of an embryo of α and in a matrix of γ.

Then the chemical driving force for the formation of this embryo is given by:

ΔG_v x volume of embryo

$= 4/3\ \pi r^3\ \Delta G_v$ for example for a spherical embryo radius r

$\equiv$ -ve below T_o

This chemical driving force will be opposed by two energy terms:

i) Strain energy, w/unit volume of embryo. This is due to the change in volume accompanying the transformation producing strain in the embryo α and surrounding matrix γ. This energy will raise the energy of the system and oppose transformation

For a given embryo shape,

Strain energy = W x volume of nucleus

$= 4/3\ \pi r^3\ W$ for a spherical embryo

$\equiv$ +ve below T_o

ii) Surface energy σ/unit of area of embryo. When the embryo is formed an interface will be created between α and γ. This interface will have a surface energy which will also raise the energy of the embryo above its surroundings and oppose transformation.

For a given nucleus shape

Surface energy = σ x surface area of nucleus

$= 4\pi r^2\ \sigma$ for a spherical nucleus

$\equiv$ +ve below T_o

Therefore we can say that the total free energy ΔG of the embryo above its surroundings is given by:

ΔG = volume of embryo x $(\Delta G_v + W)$

+ surface area of embryo x σ 5.1

$= 4/3\pi r^3(\Delta G_v + W) + 4\pi r^2\ \sigma$ 5.2

in the case of a spherical nucleus.

Note ΔG_v is -ve below T_o

The variation of ΔG with nucleus size 'a' is given in figure 5.1.

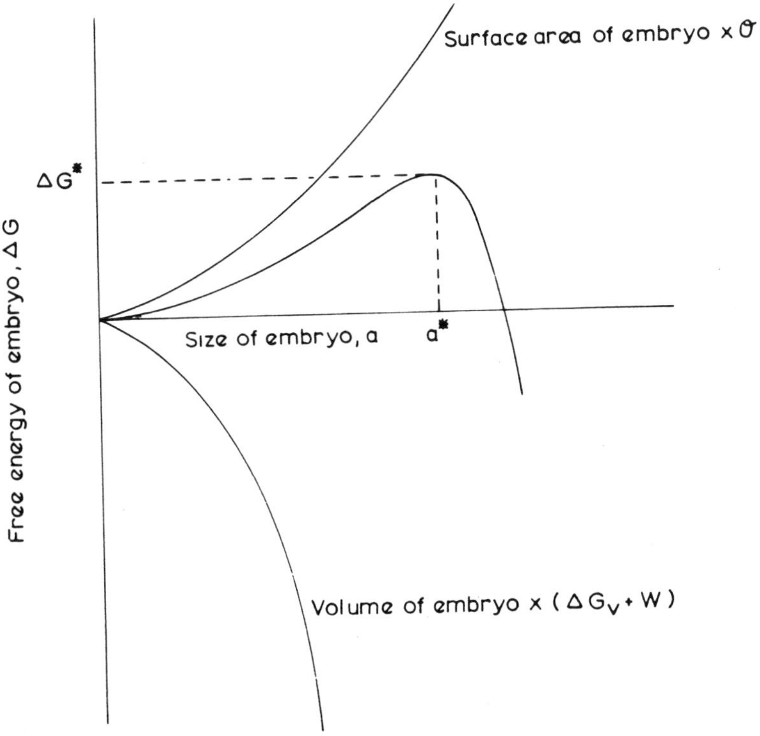

FIG. 5.1 Free energy of embryo as a function of embryo size

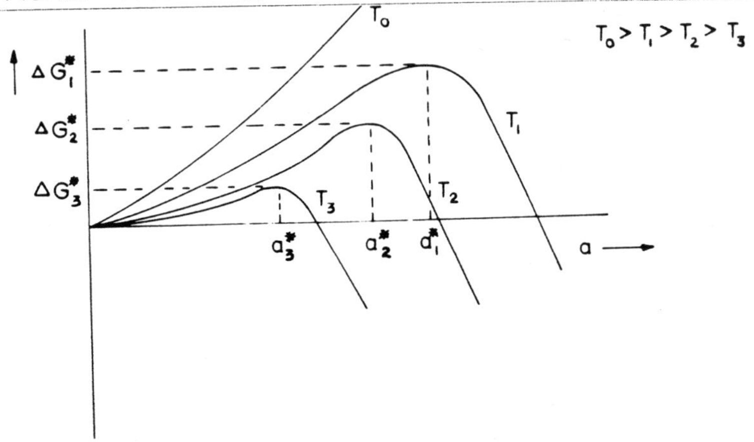

FIG. 5.2 Illustrating effect of the degree of undercooling on activation energy for nucleation ΔG^* and critical nucleus size a^*

At small values of a, the +ve term [surface area of embryo x σ]predominates while at large values of a the -ve term [volume of embryo x $(\Delta G_v + W)$] is dominant.

Thus there is a maximum value in $\Delta G = \Delta G^*$ at some critical nucleus size a*. ΔG^* is an activation barrier which has to be overcome before nucleation occurs.

At small values of a < a* an increase in embryo size, leads to an increase in free energy and therefore these embryos tend to dissolve.

For embryo sizes a > a* an increase in size leads to a decrease in energy and such embryo become stable.

a* and ΔG^* can be found as follows:

At a*, $\frac{d\Delta G}{da} = 0$

This can be solved to find a* in terms of $(\Delta G_v + W)$ and σ. Substituting a* in equation 5.1 gives ΔG^*.

Thus for a spherical embryo

$$\Delta G = 4/3\pi r^3(\Delta G_v + W) + 4\pi r^2\sigma \qquad 5.2$$

$$\therefore \quad \frac{d\Delta G}{dr} = 4\pi r^2(\Delta G_v + W) + 8\pi r\sigma = 0 \text{ at } r^*$$

$$\therefore \quad r^* = \frac{-2\sigma}{(\Delta G_v + W)} \qquad 5.3$$

Substituting r* in equation 5.2 gives

$$\Delta G^* = \frac{16}{3}\pi \frac{\sigma^3}{(\Delta G_v + W)^2} \qquad 5.4$$

Equations 5.3 and 5.4 are in agreement with qualitative argument concerning nucleation.

At small degrees of undercooling, ΔG_v is only small and therefore ΔG^* and a*(r*) are large and nucleation is difficult. While at large degrees of undercooling ΔG_v is large and therefore ΔG^* and a*(r*) are small and nucleation is easier.

This point is illustrated in figure 5.2.

5.2 Types of precipitate/matrix interface

It will be seen in equation 5.4 that $\Delta G^* \alpha\ \sigma^3$. Thus interfaces having low values of σ are preferred in nucleation.

A coherent interface has the lowest surface energy and is illustrated in figure 5.3. In this type of interface there is good matching of the precipitate and matrix planes across the interface. A nucleus invariably has one of its interfaces in this form and establishes an orientation relationship between the nucleus and matrix which is usually maintained as the nucleus grows into a precipitate. In fact in order to establish a coherent interface between nucleus and matrix, intermediate precipitates having a different crystal structure to the equilibrium phase often form first on ageing.

eg ε carbide $Fe_{2.4}C$, instead of Fe_3C on tempering of martensite

G.P zones [1] and [2] in age hardening of Al-Cu alloys

However in general there is a slight mismatch Δ between α and γ, defined by

$$\Delta = \frac{d_\gamma - d_\alpha}{d_\alpha} \qquad \text{5.5}$$

where d_α = spacing of planes of α perpendicular to interface

d_γ = spacing of planes of γ perpendicular to interface

This mismatch gives rise to strain in the α and γ phases.

To accommodate this strain, dislocation may form at the interface to produce a <u>semi-coherent interface</u>, figure 5.3.

Finally at large precipitate sizes the total strain energy accompanying the precipitate may be so large that the interface is replaced by an <u>incoherent interface</u>, figure 5.4, with a large surface energy but negligible strain energy. At an incoherent interface there is no matching of atomic planes and the atoms are disordered like in a grain boundary.

<u>Problem 5.1</u>

Derive expressions for the total free energy ΔG associated with the homogeneous creation of β nuclei from an unstable α phase for nuclei of the following shape:

(i) A disc of radius r and thickness y = 0.1r

(ii) A cylinder radius r and length y = 8r

Chemical free energy change per unit volume = ΔG_v

Energy of the α/β interface = σ/unit surface area

The β nucleus may be assumed to be incoherent with α.

What will be the activation energy and initial size of the respective nuclei? Which shape would you expect to form in preference and why?

Volume of nucleus = $\pi r^2 y$

$\therefore$ Chemical driving force for formation of nucleus = $\pi r^2 y \Delta G_v$

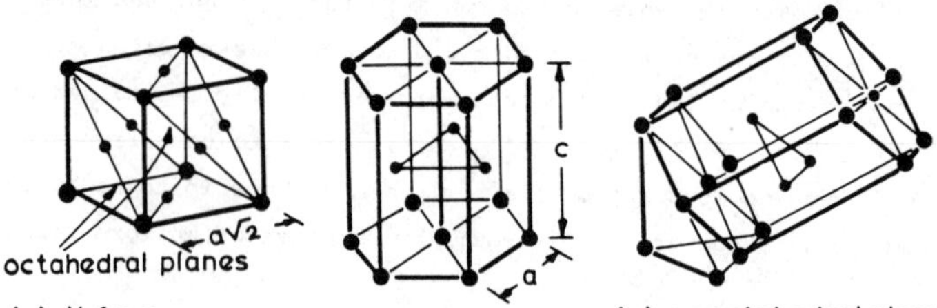

(a) γ f.c.c. face centred cubic

(b) α cph close packed hexagonal

(c) on octahedral plane of γ

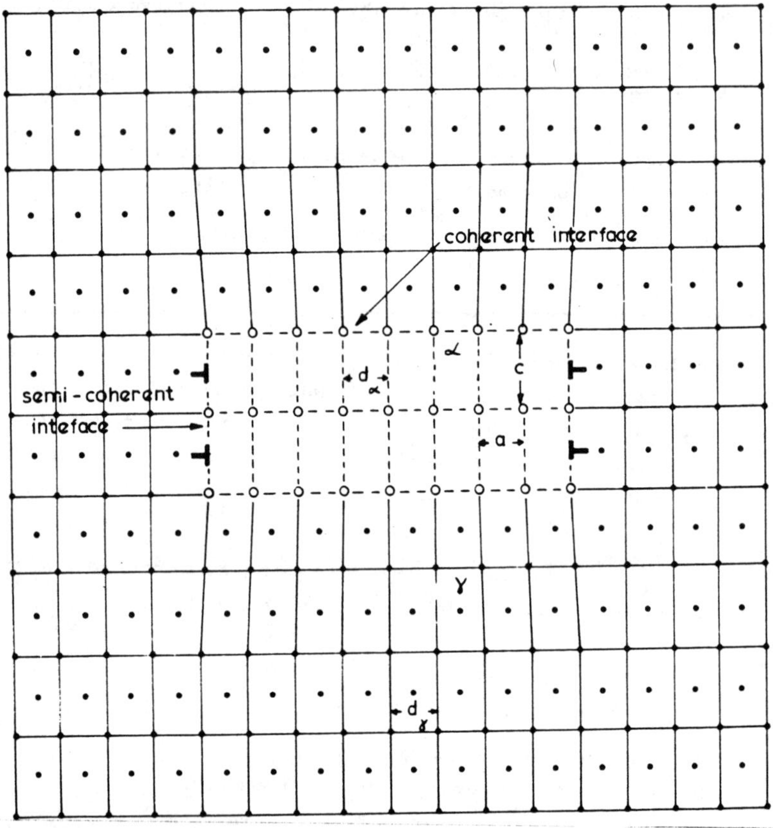

misfit strain $\delta = \dfrac{d_\gamma - d_\alpha}{d_\gamma} = 0{\cdot}05$

FIG.5.3 Illustrating coherent and semi-coherent interface.

Surface area of nucleus = $2\pi ry + 2\pi r^2$

$\therefore$ Surface energy associated with nucleus = $2\pi ry\sigma + 2\pi r^2\sigma$

Since the nucleus is incoherent the strain energy W will be zero.

Hence $\Delta G = \pi r^2 y \, \Delta G_v + 2\pi ry\sigma + 2\pi r\sigma$

$= \pi r^2 y \, \Delta G_v + 2\pi(ry + r^2)\sigma.$

In general y = ar where a = 0.1 for disc and 8 for cylinder.

$\therefore \Delta G = \pi a r^3 \, \Delta G_v + 2\pi r^2 (a + 1)\sigma$

$$\frac{\partial \Delta G}{\partial r} = 3\pi a r^2 \, \Delta G_v + 4\pi r(a + 1)\sigma = 0 \text{ at } r = r^*$$

$\therefore$ Critical nucleus size $r^* = -\dfrac{4(a + 1)\sigma}{3a \, \Delta G_v}$

Substituting for r = r* back in ΔG gives

Activation energy for nucleation ΔG*

$$= -\pi a \frac{64(a + 1)^3\sigma^3\Delta G_v}{27a^3 \, \Delta G_v{}^3} + 2\pi \frac{16(a + 1)^2\sigma^2(a + 1)\sigma}{9a^2 \, \Delta G_v{}^2}$$

$$= -\frac{64\pi}{27} \frac{(a + 1)^3}{a^2} \frac{\sigma^3}{\Delta G_v{}^2} + \frac{32\pi}{9} \frac{(a + 1)^3}{a^2} \frac{\sigma^3}{\Delta G_v{}^2}$$

$$= \frac{32\pi}{27} \frac{(a + 1)^3}{a^2} \frac{\sigma^3}{\Delta G_v{}^2}$$

(i) For disc a = 0.1

$$\therefore r^* = -\frac{4.4}{0.3} \frac{\sigma}{\Delta G_v} = -\frac{4.4}{3} \frac{\sigma}{\Delta G_v}$$

$$\Delta G^* = \frac{32\pi}{27} \frac{(1.1)^3}{0.01} \frac{\sigma^3}{\Delta G_v{}^2}$$

$$= 157.75\pi \frac{\sigma^3}{\Delta G_v{}^2}$$

(ii) For cylinder a = 8

$$\therefore r^* = \frac{-4 \times 9}{24} \frac{\sigma}{\Delta G_v} = -\frac{3}{2} \frac{\sigma}{\Delta G_v}$$

$$\Delta G^* = \frac{32\pi}{27} \frac{9 \times 81}{64} \frac{\sigma^3}{\Delta G_v{}^2}$$

$$= 13.5\pi \frac{\sigma^3}{\Delta G_v{}^2}$$

Hence the cylinder will form in preference to the disc because of its lower activation energy, this is because it has a larger volume to surface ratio (4r/9) than the disc (r/22).

5.3 The metastable equilibrium concentration of embryoes

Consider embryos of total free energy ΔG_n containing n atoms.

These form on possible N sites/unit volume of γ.

If the concentration of embryos of size n is C_n, then we can write the reaction for the formation of these as

$$N \rightarrow C_n$$

with free energy ΔG_n accompanying the reaction.

The equilibrium constant K for this reaction is given by

$$K = \frac{\text{concentration of products}}{\text{concentration of reactants}} = \frac{C_n}{N}$$

where $\Delta G_n = -kT \ln K$

$$= -kT \ln \frac{C_n}{N}$$

ie $C_n = N \exp\left(-\frac{\Delta G_n}{kT}\right)$ 5.6

Thus the concentration of critical nuclei will be given by

$C^* = N \exp\left(-\frac{\Delta G^*}{kT}\right)$ 5.7

5.4 The Classical Nucleation Rate

In the classical nucleation theory due to Volmer and Weber (5.1) it is assumed that:

Rate of Nucleation/unit volume I

= No of critical nuclei/unit volume x Rate of growth of each nucleus β^*

$= N \exp - \frac{\Delta G^*}{kT} \times \beta^*$

Now β^* = No of atoms on surface of initial nucleus x rate at which atoms join nucleus

$= S^* \times N_\gamma^{2/3} \times \nu \times p$

where S^* = surface area of critical nucleus

N_γ = No of atoms/unit volume of γ

ν = frequency of vibration of atoms

p = probability that atom is moving in the right direction to join the nucleus

Diffusion theory gives us:

$p\nu = \frac{D}{a_\gamma^2} = \frac{D_0}{a_\gamma^2} \exp - \frac{Q}{kT}$ 5.9

where D = appropriate diffusion coefficient for atoms joining the nucleus

Q = activation energy for diffusion of atoms joining the nucleus

a_γ = lattice parameter of γ

Hence we have

$$I = NS^* N_\gamma^{2/3} \frac{D_o}{a_\gamma^2} \exp - \frac{\Delta G^*}{kT} \exp - \frac{Q}{kT} \quad \ldots\ldots\ldots\ldots\ldots\ldots\ldots\ldots\ldots \quad 5.10$$

Now S* will vary with ΔG*, which in turn will vary with temperature through ΔG_v. However this temperature dependence of S* can be neglected compared with the temperature dependence of the exponential terms and we can write

$$I = A \exp - \frac{\Delta G^*}{kT} \exp - \frac{Q}{kT}$$

$$= A \exp - \frac{(\Delta G^* + Q)}{kT} \quad \ldots\ldots\ldots\ldots\ldots\ldots\ldots\ldots\ldots\ldots\ldots\ldots \quad 5.11$$

The temperature dependence of I is shown schematically in figure 5.5, with the corresponding TTT curve in figure 5.6. It will be seen that this theory predicts the well established 'C' curve behaviour of transformations in TTT diagrams. At small degrees of undercooling nucleation is difficult although diffusion (ie growth of the critical nuclei) is fast. Hence the overall rate of nucleation is controlled by the slowest process, nucleation

ie $\exp - \frac{\Delta G^*}{kT}$.

At large degrees of undercooling although nucleation is easy (ΔG* small) diffusion or growth is sluggish and therefore the overall rate of nucleation is controlled by the rate of growth, ie $\exp - \frac{Q}{kT}$.

Problem 5.2 Q.1 Associateship in Metallurgy, Physical Metallurgy, June 1973, Sheffield Polytechnic.

The isothermal transformation of austenite to ferrite in an Fe-8.5%Cr alloy was followed by a magnetic method at several temperatures. The following times for 5% transformation were obtained at the given isothermal temperatures:

Isothermal Temperature °C	677	657	637	617	607	587	567
Time to 5% transformation (minutes)	1.5	2.0	4.0	9.0	15	45	119

For this transformation, calculate:

(a) The activation energy for growth.

(b) The activation energy for nucleation at 677°C.

Boltzmann's constant, $k = 1.38 \times 10^{-23}$ JK^{-1}

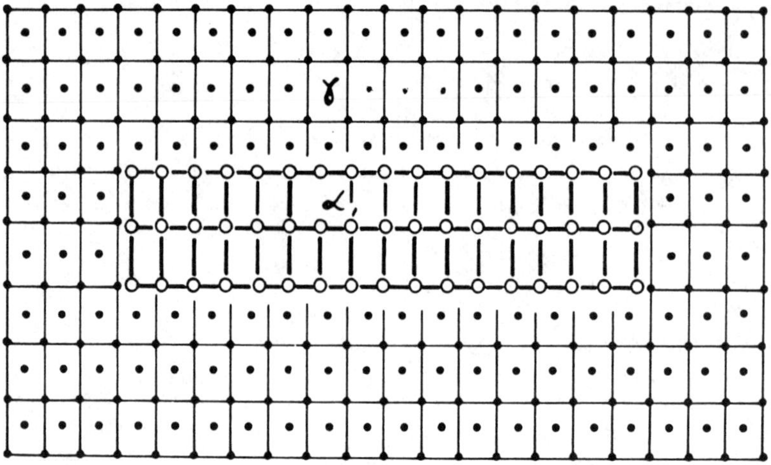

FIG. 5.4 Illustrating incoherent interface.

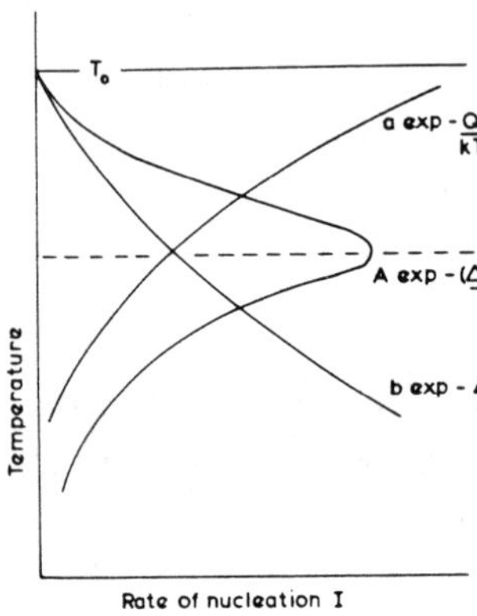

FIG. 5.5 Temperature dependence of nuclation rate.

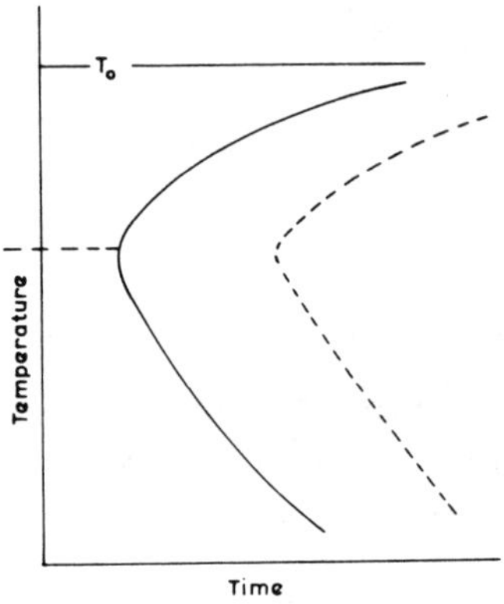

FIG. 5.6 Schematic TTT diagram for transformation occurring in figure 5.3.

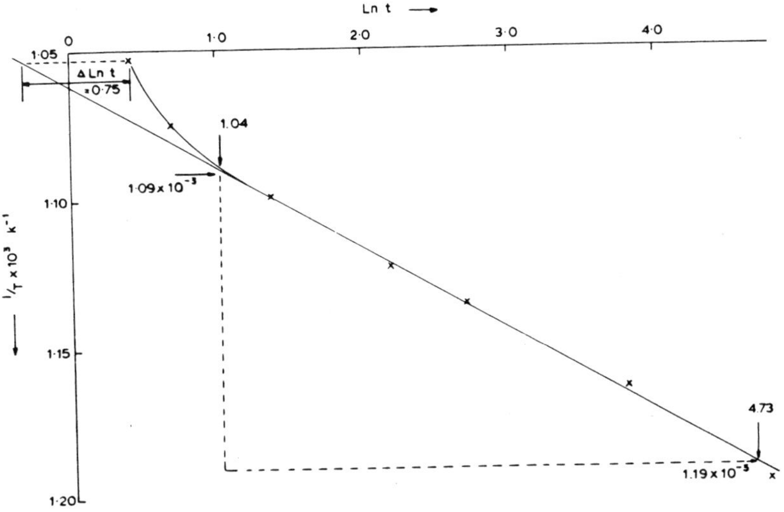

Fig. 5.7 Problem 5.2 Determination of activation energy for growth and activation energy for nucleation in an Fe 8.5%Cr alloy.

Temp ^{o}C	677	657	637	617	607	587	567
677 K	950	930	910	890	880	860	840
$^{1}/_{T} \times 10^{3}$	1.053	1.075	1.099	1.123	1.136	1.163	1.19
t	1.5	2.0	4.0	9.0	15	45	119
lnt	0.405	0.693	1.386	2.197	2.708	3.807	4.279

Assuming rate of transformation $\alpha \frac{1}{t}$, we have

$$\frac{1}{t} = A \exp - \frac{(\Delta G^* + Q)}{kT}$$

$$\text{i.e. } -\ln t = \ln A - \frac{(\Delta G^* + Q)}{kT}$$

$$\text{i.e. } \ln t = \frac{Q}{k} \times \frac{1}{T} + \frac{\Delta G^*}{k} \ \frac{1}{T} - \ln A \ \dots\dots\dots\dots \quad (1)$$

At large degrees of undercooling, below the nose of the 'C' curve, $\Delta G^* << Q$, therefore

$$\ln t = \frac{Q}{k}\frac{1}{T} - \ln A \ \dots\dots\dots\dots \quad (2)$$

∴ From the Arrhenius plot, figure 5.7 slope of graph

$$= \frac{\Delta \ln t}{\Delta 1/T} = \frac{4.73 - 1.04}{(1.19 - 1.09) \times 10^{-3}} = \frac{Q}{k}$$

∴ $Q = \underline{\underline{5.1 \times 10^{-19} \text{ J/atom}}}$

At temperature T_1, subtracting equation (2) from (1)

$$\Delta \ln t = \frac{\Delta G^*}{k} \times \frac{1}{T_1}$$

At $677^{o}C$ = 950K $\quad \Delta \ln t = 0.75 = \dfrac{\Delta G^*}{1.38 \times 10^{-23}} \times \dfrac{1}{950}$

∴ $\Delta G^* = \underline{\underline{9.8 \times 10^{-21} J}}$

In actual fact this particular transformation is the equi-axed ferrite transformation which occurs by the massive transformation mechanism. Since the growth of the ferrite by this mechanism is rapid, the kinetics of the transformation is controlled by nucleation and not nucleation and growth. K.C. Russell's theory of incubation periods applies to the transformation - see Chapter 9.

5.5 Evaluation of Strain Energy W

Let Δ = linear dilatation accompanying the formation of α from γ

$$= \frac{dl}{l}$$

$$= 1/3 \frac{\Delta V}{V} \ \dots\dots\dots\dots \quad 5.12$$

Since $V = \alpha l^3$ where α = shape factors

$\therefore \quad dV = 3\alpha l^3 \quad \therefore \quad \frac{dl}{l} = \frac{1}{3}\frac{\Delta V}{V}$

Then Nabarro (5.2) has shown that for insertion of a rigid sphere in an elastic matrix

$$W = 6G\Delta^2 \qquad 5.13$$

where G = shear modulus of matrix.

Nabarro used this estimate to show that for incoherent precipitates that strain energy was a minimum for a disc.

Alternatively Crum cited in reference (52) showed that for insertion into a cavity in the matrix, a body of the same shape and rigidity as the matrix but different size whatever the shape of the body

$$W = \frac{6G\Delta^2}{(1 + 4G/3K)} = 2G\Delta^2\frac{(1 + R)}{(1 - R)} = 4G\Delta^2 \qquad 5.14$$

assuming R = Poisson's ratio = 1/3

K = Bulk modulus

For insertion of an elastic sphere in a rigid matrix problem 1.3

$$W = 4.5K\Delta^2 = 12G\Delta^2 \text{ with Poisson's ratio} = 1/3 \qquad 5.15$$

For coherent precipitates Δ can be obtained from the planar spacing of the atoms at the matrix/precipitate interface as defined before

$$\text{i.e.} \quad \Delta = \frac{d_\gamma - d_\alpha}{d_\alpha} \qquad 5.16$$

More sophisticated calculations can use the approach of Eshelby (5.3).

5.6 Effective metastable equilibrium temperature T_o'

Equation 5.4 showed that for homogeneous nucleation of a sphere

$$\Delta G^* = \frac{16}{3}\pi\frac{\sigma^3}{(\Delta G_v + W)^2} \qquad 5.4$$

Now $\Delta G^* \rightarrow \infty$ when $\Delta G_v + W = 0$, i.e. we can define an effective metastable equilibrium temperature T_o' in the presence of strain energy such that

$$\Delta G_v + W = 0 \qquad 5.17$$

Thus W opposes ΔG_v and thus transformation can only occur below T_o' when there is strain in the precipitate. This will be true whatever the shape of the precipitate since in the formula for the activation energy for nucleation, ΔG^*, the term $(\Delta G_v + W)$ will appear in the denominator.

Note it is not true to say that an increase in the effective surface energy term σ will lower T_o as suggested by some authors as this term appears in the numerator for ΔG^* and does not directly oppose ΔG_v. However an increase in σ will increase ΔG^* and thus slow down the nucleation rate and apparently lower T_o.

5.7 Equilibrium shape of precipitates

It is generally found that precipitates tend to grow in the direction of least mismatch Δ (5.4). This is because this form of growth will minimise strain energy W and maximise the coherent surface area. Δ can be obtained from a knowledge of the lattice parameter of the matrix and precipitate and the orientation relationship between precipitate and matrix.

Problem 5.3 Q.4 Associateship in Metallurgy, Physical Metallurgy 1, June 1972, Sheffield Polytechnic.

Define the percentage mismatch between a precipitate and its matrix, and indicate the importance of this mismatch in determining the shape and morphology of precipitates.

By considering the appropriate mismatch values show that it is highly likely that:

(a) V_4C_3 will form as a square plate on $\{100\}_\alpha$ in ferrite.
The orientation relationship is such that
$(100)\alpha//(100)V_4C_3$
$(011)\alpha//(010)V_4C_3$
$(01\bar{1})\alpha//(00\bar{1})V_4C_3$
Lattice parameter of fcc V_4C_3 = 4.16Å

and

(b) Fe_3C will form as a lath with <111>α growth direction and {110}α habit plane in ferrite. The orientation relationship is such that
$(1\bar{1}0)\alpha//(100)Fe_3C$
$(111)\alpha//(010)Fe_3C$
$(11\bar{2})\alpha//(001)Fe_3C$
Lattice parameter of orthorhombic Fe_3C
a_o = 4.52Å; b_o = 5.09Å; c_o = 6.74Å
Lattice parameter of αFe = 2.86Å.

(a) For V_4C_3 in ferrite

$$\Delta \text{ in } [100]_\alpha = \frac{d_{(100)}(V_4C_3) - d_{(100)}(\alpha)}{d_{(100)}(V_4C_3)}$$

$$= \frac{4.16 - 2 \times 2.86}{4.16} = \underline{\underline{37.5\%}}$$

$$\Delta \text{ in } [011]_{\alpha} = \frac{d_{(010)}(V_4C_3) - d_{(011)}(\alpha)}{d_{(010)}(V_4C_3)}$$

$$= \frac{4.16 - 2.86\sqrt{2}}{4.16} = \underline{\underline{2.8\%}}$$

$$\Delta \text{ in } [01\bar{1}]_{\alpha} = \frac{d_{(00\bar{1})}(V_4C_3) - d_{(01\bar{1})}(\alpha)}{d_{(00\bar{1})}(V_4C_3)}$$

$$= \frac{4.16 - 2.86\sqrt{2}}{4.16} = \underline{\underline{2.8\%}}$$

$$= \underline{\underline{2.8\%}}$$

Δ is very large in the $[100]_{\alpha}$ direction, and hence V_4C_3 will not grow in this direction.

The mis-match Δ is small in $(011)_{\alpha}$ and $(01\bar{1})_{\alpha}$ directions, hence V_4C_3 will grow in these directions.

i.e. V_4C_3 grows as a square plate on $(100)_{\alpha}$.

(b) For Fe_3C in ferrite

$$\Delta \text{ in } (1\bar{1}0)_{\alpha} = \frac{d_{(100)}(Fe_3C) - d_{(1\bar{1}0)}(\alpha)}{d_{(100)}(Fe_3C)}$$

$$= \frac{4.52 - 2.86\sqrt{2}}{4.52} = \underline{\underline{+10.5\%}}$$

$$\Delta \text{ in } (111)_{\alpha} = \frac{d_{(010)}(Fe_3C) - d_{(111)}(\alpha)}{d_{(010)}(Fe_3C)}$$

$$= \frac{5.09 - 2.86\sqrt{3}}{5.09} = \underline{\underline{2.7\%}}$$

$$\Delta \text{ in } (11\bar{2})_{\alpha} = \frac{d_{(001)}(Fe_3C) - d_{(11\bar{2})}(\alpha)}{d_{(001)}(Fe_3C)}$$

$$= \frac{6.74 - 2.86\sqrt{5}}{6.74} = \underline{\underline{5.1\%}}$$

Mismatch Δ is a maximum in $(1\bar{1}0)_{\alpha}$ direction, hence no growth occurs in this direction.

Δ is a minimum in $[111]_\alpha$ direction, hence maximum growth occurs in this direction.

Δ is only 5.1% in $[11\bar{2}]_\alpha$ direction, so growth occurs on plane containing $[111]_\alpha$ = $[11\bar{2}]_\alpha$ i.e. $[1\bar{1}0]_\alpha$.

i.e. growth occurs as a lath on $\{110\}_\alpha$ plane in $\langle 111 \rangle_\alpha$ direction.

<u>Problem 5.4</u> Q.5 Associateship in Metallurgy, Physical Metallurgy, 1 June 1976, Sheffield Polytechnic.

Define the mismatch parameter, δ, between a precipitate and its matrix, and state the importance of this mismatch in determining the shape of precipitates.

Determine the shape of a f.c.c. precipitate γ^1, lattice parameter 3.60Å, which will form in austenite of lattice parameter 3.58Å. The orientation relationship is such that:-

$(100)_\gamma // (100)_{\gamma'}$

$(010)_\gamma // (010)_{\gamma'}$

$(001)_\gamma // (001)_{\gamma'}$

Calculate the critical nucleus size for precipitation of coherent particles of γ' in austenite at 943K from the following data:-

Surface energy of coherent γ'/austenite interface = 5×10^{-8} Jmm^{-2}.

Latent heat of transformation per unit volume of γ' (ΔH_v) = -2.25×10^{-1} Jmm^{-3}.

The $\gamma/(\gamma + \gamma^1)$ solvus temperature T_o = 1143K.

Shear modulus of austenite, G = 57 kN mm^{-2}.

Strain energy per unit volume of γ' for a coherent precipitate = $6G\delta^2$.

<u>References</u>

5.1 M. Volmer and A. Weber, Phy. Chem, 1925, <u>119</u> p.277.

5.2 F.R.N. Nabarro, Proc. Roy. Soc., 1940, Vol 175A, pp.519-538.

5.3 J.B. Eshelby, Proc. Roy. Soc, 1957, Vol. A241, pp. 376-396.

5.4 J.A.Whiteman and S.R.Keown, Sheff. Univ.Met.SocJ, 1971, <u>10</u>, p.30.

6. Heterogeneous Nucleation

When homogeneous nucleation theory is applied to the solidification of liquids it is found that the theoretical undercooling ΔT required for solidification is of the order of 200K. In practice undercooling for solidification of liquids are $\sim$2-5K.

This is due to heterogeneous nucleation of solids on foreign surfaces effectively lowering the surface energy σ in equation 5.1 and making nucleation much more easier.

In a classical experiment by Turnbull on the freezing of Hg and its alloys he divided the liquid into a large number of drops (10-50μ) suspended in oil or slag, so that the number of heterogeneous sites was reduced. Turnbull counted the number of droplets freezing as a function of undercooling and obtained the results shown in figure 6.1.

Some droplets were observed to freeze at small degrees of undercooling (i.e. 1). This was due to heterogeneous nucleation in these droplets. Other droplets required a large degree of undercooling (i.e. 3), in which homogeneous nucleation occurred. Some intermediate undercooling (2) was also obtained due to less effective catalysts for heterogeneous nucleation.

Simple heterogeneous nucleation theory for solidification

The following treatment although not strictly correct serves to illustrate the effect of wetting and contact angle in heterogeneous nucleation.

Consider the nucleation of a solid α from liquid L on a foreign surface or substrate β, figure 6.2.

Resolving surface energies horizontally and vertically

$$\sigma_{L\beta} = \sigma_{\alpha\beta} + \sigma_{L\alpha}\cos\theta \quad \text{.......} \quad 6.1$$

$$\sigma_{T} = \sigma_{L\alpha}\sin\theta \quad \text{.......} \quad 6.2$$

Let $A_{L\alpha}$ = surface area of upper surface between α and L.

When the nucleus is formed a β/liquid surface of radius l is destroyed.

Hence the change in surface energy in forming the nucleus =

$$= A_{L\alpha}\sigma_{L\alpha} + \pi l^2\sigma_{\alpha\beta} - \pi i^2\sigma_{L\beta}$$

$$= A_{L\alpha}\sigma_{L\alpha} + \pi l^2(\sigma_{\alpha\beta} - \sigma_{L\beta})$$

$$= A_{L\alpha}\sigma_{L\alpha} + \pi l^2(-\sigma_{L\alpha}\cos\theta) \quad \text{from equation 6.1}$$

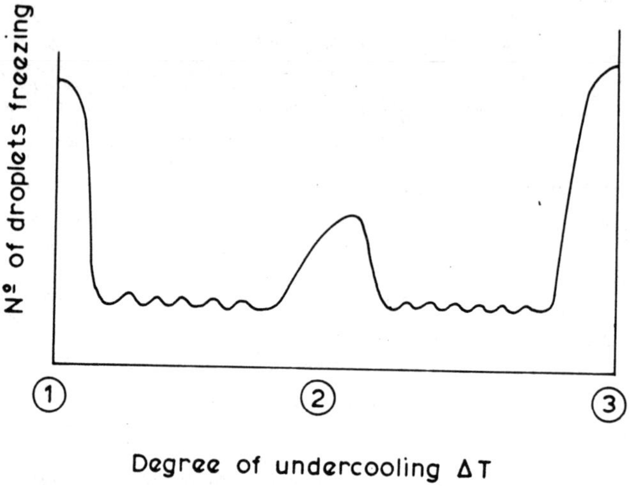

FIG. 6.1 Freezing of droplets as a function of degree of undercooling, after D. Turnbull (6.1)

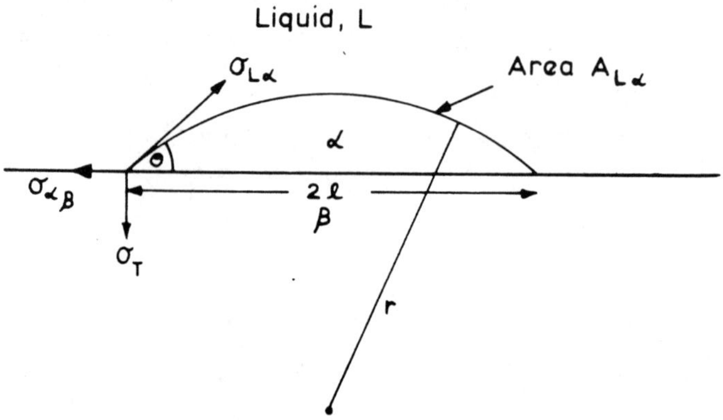

FIG. 6.2 Illustrating hetrogeneous nucleation from the liquid on a catalyst β.

$= \sigma_{L\alpha}(A_{L\alpha} - \pi l^2\cos\theta)$ 6.3

Without a catalyst the change in surface energy on forming the 'nucleus'

$= \sigma_{L\alpha}(A_{L\alpha} + \pi l^2)$ 6.4

Equation 6.4 is of course physically incorrect since in the absence of a catalyst homogeneous nucleation will occur and the nucleus will 'ball-up' into a sphere.

Nevertheless equations 6.3 and 6.4 serve to illustrate the effect of contact angle θ on nucleation.

If $\theta = 180^\circ$ solid α does not wet the particle β - point contact.

$\cos\theta = -1$ and change in surface energy for heterogeneous nucleation

$= \sigma_{L\alpha}(A_{L\alpha} - \pi l^2)$

≡ same as equation 6.4 for homogeneous nucleation.

ie particle is ineffective as catalyst.

If $\theta = 0^\circ$ i.e. perfect wetting

$\cos\theta = 1$ and change in surface energy for heterogeneous nucleation

$= \sigma_{L\alpha}(A_{L\alpha} - \pi l^2)$

$= 0$ since $\theta = 0$ and $A_{L\alpha} = \pi l^2$

In this case there is no barrier to nucleation and α spreads right across β as a thin layer.

This situation is realised in practice when the catalyst β is in fact solid α.

For angle of θ between 0 and 180 the change in surface energy for heterogeneous nucleation will be less than given by equation 6.4.

In conclusion we can see that nucleation is easier on a catalyst and the effectiveness of the catalyst depends upon the contact angle θ.

Problem 6.1

Determine the degree of undercooling for homogeneous solidification of pure copper. You may assume the solid/liquid interfacial energy is ∿200 ergs/cm^2

Melting point of copper = 1352 K

Latent heat of fusion of copper = 13.05 kJ/mol

Relative atomic mass of copper = 63.54

Density of copper = 8.93 g/cc

Boltzmann constant $k = 1.38 \times 10^{-23}$ Jdeg^{-1}

For homogeneous nucleation of a sphere, section 5.1,

$$\Delta G^* = \frac{16}{3} \pi \frac{\sigma^3}{\Delta G_v^2} \quad \ldots\ldots \quad 6.5$$

The strain energy W accompanying solidification will be negligible as the volume change will easily be accommodated by the liquid.

Now $\Delta G_v = \Delta H_v - T\Delta S_v$

At the melting temperature T_M, $\Delta G_v = 0$

$\therefore \quad \Delta H_v = T_M \Delta S_v$

Hence for a degree of undercooling ΔT,

$$\Delta G_v = \frac{\Delta H_v \Delta T}{T_n} \quad \ldots\ldots \quad 6.6$$

The number of critical nuclei n_c/unit volume will be given by Boltzmann statistics as

$$n_c = N \exp - \frac{\Delta G^*}{kT}$$

$$= N \exp - \frac{16}{3} \pi \frac{\sigma^3}{\Delta G_v^2}$$

$$= N \exp - \frac{16}{3} \pi \frac{\sigma^3 T_M^2}{\Delta H_v^2 \Delta T^2 kT} \quad \ldots\ldots \quad 6.7$$

from equations 6.5 and 6.6

N is the number of nucleation sites/unit volume $\sim$ Avogadro's number $\sim 10^{23}$.

We will assume that solidification occurs when $n_c \sim 1$.

$T_M = 1352$ K

$\Delta H_v = 13.05$ kJ/mol $= 13.05/63.54$ kJ/g

$= 13.05/63.54 \times 8.93$ kJ/cc $= 1.834$ kJ/cc

$\sigma = 200$ ergs/cm^2 $= 2 \times 10^{-5}$ J/cm^2 $\qquad$ 1 erg $= 10^{-7}$ J

Taking logarithms of equation 6.7

$$\ln \frac{n_c}{N} = - \frac{16}{3} \pi \frac{\sigma^3 T_M^2}{\Delta H_v^2 \Delta T^2 kT}$$

ie $$\ln \frac{N}{n_c} = \frac{16}{3} \pi \frac{\sigma^3 T_M^2}{\Delta H_v^2 \Delta T^2 kT}$$

$$\ln 10^{23} = \frac{16}{3} \pi \frac{(2 \times 10^{-5})^3 (1352)^2}{(1834)^2 \times 1.38 \times 10^{-23}\, T\Delta T^2}$$

$\therefore \quad T\Delta T^2 = 9.97 \times 10^7$

$\Delta T = 309^oC$

6.2 Heterogeneous nucleation in solids

This is the type of nucleation observed in practice. Homogeneous nucleation is a rare event only occurring at high degrees of supersaturation in a few systems.

eg Cu - Co and spinodal decomposition.

The free energy accompanying heterogeneous nucleation on a defect is given by

ΔG = volume of nucleus (ΔG_v + W) + area of nucleus x σ - area destroyed x σ_d

The energy associated with the defect, (area destroyed) x σ_d leads to a decrease in free energy, ie a decrease in the surface energy term.

Hence the activation energy for nucleation will depend upon the energy of the defect surface destroyed and will be least when this energy (area x σ_d) is greatest.

Thus we can say for activation energy ΔG^*

$$\Delta G^*_{\text{vacancy cluster}} > \Delta G^*_{\text{dislocation}} > \Delta G^*_{\text{grain boundary}} > \Delta G^*_{\text{free surface}}$$

The rate of hetrogeneous nucleation is given by

$$I_{het} = AN_s \exp - \frac{\Delta G^*_{het}}{RT} \quad \ldots\ldots\ldots\ldots\ldots\ldots\ldots\ldots \quad 6.8$$

Where A = constant dimensions time^{-1}

N_s = No of hetrogeneous sites.

Note the rate of nucleation depends upon the number of sites and these will be limited compared with the number of homogeneous sites. Thus in order to allow for the reduced number of precipitation sites and for hetrogeneous nucleation to occur rather than homogeneous, typically (6.2)

$$\Delta G^*_{het} < 0.6\ \Delta G^*_{hom}$$

Problem 6.1 Show that for hetrogeneous nucleation of a sphere on a grain boundary

$$\Delta G^*_{het} = \frac{27}{64}\ \Delta G^*_{hom}$$

You may assume that the surface energy of the grain boundary $\sigma_{g.b}$ is equal to the surface energy of the nucleus/parent interface.

The physical situation is shown in figure 6.3.

The free energy ΔG_{het} accompanying the formation of this spherical nucleus is given by

$$\Delta G_{het} = \frac{4}{3}\pi r^3 \Delta G_v + 4\pi r^2\sigma - \pi r^2\sigma_{g.b} \quad \text{.......................} \quad 6.9$$

Note the strain energy W accompanying this nucleus will be negligible as the volume change in forming the nucleus will be easily accommodated by the grain boundary.

Differentiating with respect to r gives

$$\frac{d\Delta G_{het}}{dr} = 4\pi r^2 \Delta G_v + 8\pi r\sigma - 2\pi r\sigma_{g.b}$$

$$= 4\pi r^2 \Delta G_v + 2\pi r(4\sigma - \sigma_{g.b})$$

Hence at critical nucleus size r^*, $\frac{d\Delta G}{dr} = 0$

and $$r^* = -\frac{2\pi(4\sigma - \sigma_{g.b})}{4\pi\Delta G_v} = -\frac{(4\sigma - \sigma_{g.b})}{2\Delta G_v} \quad \text{........................} \quad 6.10$$

Substituting back in equation 6.9 gives

$$\Delta G^*_{het} = -\frac{4}{3}\pi\frac{(4\sigma - \sigma_{g.b})^3}{8\Delta G_v^3}\Delta G_v + 4\pi\frac{(4\sigma - \sigma_{g.b})^2}{4\Delta G_v^2}\sigma$$

$$- \pi\frac{(4\sigma - \sigma_{g.b})^2}{4\Delta G_v^2}\sigma_{g.b}$$

$$= -\frac{\pi}{6}\frac{(4\sigma - \sigma_{g.b})^3}{\Delta G_v^2} + \pi\frac{(4\sigma - \sigma_{g.b})^3}{4\Delta G_v^2}$$

$$= \frac{\pi}{12}\frac{(4\sigma - \sigma_{g.b})^3}{\Delta G_v^2}$$

If $\sigma = \sigma_{g.b}$, then $$\Delta G^*_{het} = \frac{9}{4}\pi\frac{\sigma^3}{\Delta G_v^2} \quad \text{..........................} \quad 6.11$$

For homogeneous nucleation, (section 5.1)

$$\Delta G^*_{hom} = \frac{16}{3}\pi\frac{\sigma^3}{\Delta G_v^2} \quad \text{..................................} \quad 6.12$$

Hence we have, comparing equations 6.11 and 6.12.

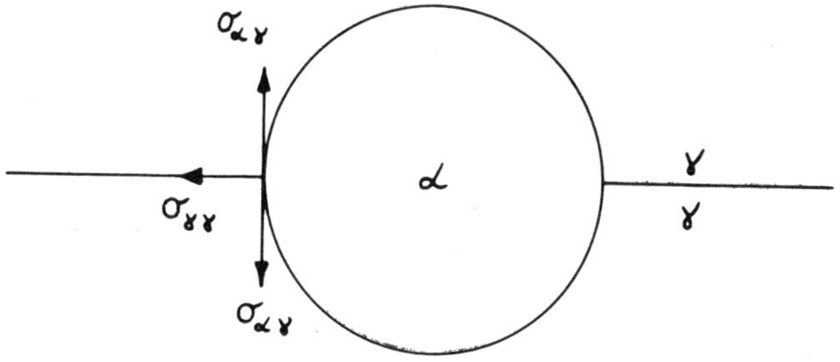

FIG. 6.3 Illustrating hetrogeneous nucleation of a sphere on a grain boundary.

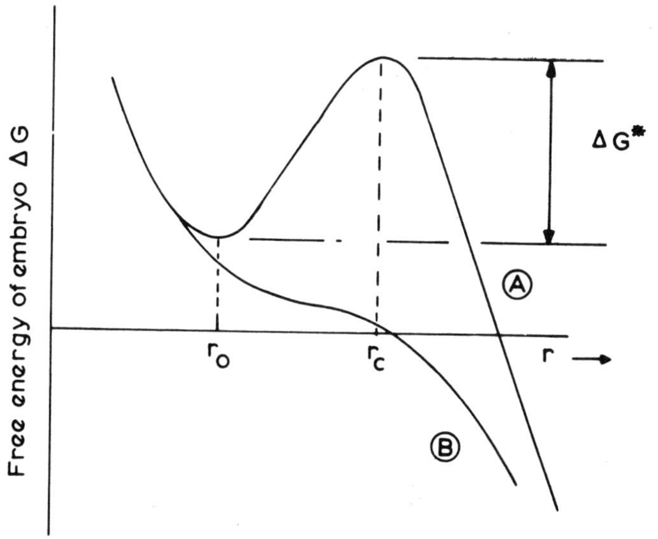

FIG. 6.4 Variation of free energy of embryo ΔG, with size r for precipitation on dislocation.

$$\Delta G^*_{het} = \frac{9}{4}\pi \times \frac{3}{16\pi}\Delta G^*_{hom}$$

$$= \frac{27}{64}\Delta G^*_{hom}$$

Further reductions in ΔG^* is accomplished at junctions between 3 and 4 grains. Clemm and Fisher (6.3) showed that the activation energy for nucleation at 4 grain junctions is about $\frac{1}{2,000}$ that for homogeneous nucleation.

Problem 6.2 BSc (Hons) finals in Metallurgy, Physical Metallurgy, June 1975, Sheffield Polytechnic, Department of Metallurgy, question 4.

Derive the conditions for there being no thermodynamic barrier to nucleation of precipitates on dislocations assuming the strain energy per unit length of a dislocation line at a distance r is given by $A \log_e r$ where A is an elastic constant.

Calculate the chemical driving force ΔG_v, and the degree of undercooling for the precipitation of Fe_3C on dislocations in αFe containing 0.01%C.

DATA:
For αFe A is 5×10^{-10}N.
Energy (α) of Fe_3C/α interface is $0.2 J/m^2$.
Latent heat of transformation per unit volume (ΔH_v) of Fe_3C is $1.7\ J/mm^3$.
The $\alpha/\alpha + Fe_3C$ solvus temperature (T_o) at 0.01%C is 900 K.

This is a simple model due to Cahn (6.5). Change in free energy/unit length of dislocation for a cylindrical nucleus radius r is given by

$$\Delta G = \pi r^2 \times 1 \times \Delta G_v + 2\pi r \times 1 \times \sigma - A\ln r \quad \text{.................. 6.12}$$

Alnr comes from dislocation theory

$$E = \frac{Gb^2}{4\pi}\ln\left(\frac{r_1}{r_o}\right)$$

$$= A\ln r - A\ln r_o$$

Differentiating equation 6.12 with respect to r, $\frac{d\Delta s}{dF} = 2\pi r\Delta G_v + 2\pi\sigma - \frac{A}{r}$ and setting equal to zero to find the critical nucleus size gives a quadratic in r

$$2\pi r^2\Delta G_v + 2\pi r\sigma - A = 0$$

cf $\quad ax^2 + bx + c = 0$

Solution $x = \dfrac{-b \pm \sqrt{b^2 - 4ac}}{2a}$

Therefore there are two solutions

$$r = \frac{-2\pi\sigma \pm \sqrt{4\pi^2\sigma^2 + 4 \times 2\pi\Delta G_V A}}{2 \times 2\pi\Delta G_V}$$

$$= \frac{-\sigma}{2\Delta G_V}\left[1 \pm \sqrt{1 + \frac{2A\Delta G_V}{\sigma\pi^2}}\right] \quad \ldots\ldots\ldots\ldots\ldots\ldots \quad 6.11$$

The negative sign arises because ΔG_V is negative. The variation of ΔG with r is shown in figure 6.4. Note to find the activation energy for nucleation ΔG^*, the total free energy change at r_o and r_c is evaluated and the difference between these quantities gives ΔG^*

ie $\quad \Delta G^*_{dislocation} = \Delta G|_{r_c} - \Delta G|_{r_o}$

Note to obtain real roots of the quadratic we must have $|2A\Delta G_V| < \pi\sigma^2$.

If $|2A\Delta G_V| > \pi\sigma^2$ there is neither a minimum nor a maximum and there is no barrier to nucleation curve B, figure 6.4.

$$\Delta G_V = \Delta H_V - T\Delta S_V$$

At $\quad T_o = 900K,\ \Delta G_V = 0 \qquad \therefore\ \Delta H_V = T_o\Delta S_V$

$$\therefore \quad \Delta G_V = \Delta H_V - \frac{T\Delta H_V}{T_o} = \frac{\Delta H_V(T_o - T)}{T_o}$$

$$= \frac{1.7}{900}\Delta T\ J/mm^3 = \frac{1.7}{900}\Delta T \times 10^{+9}\ J/m^3$$

$\therefore$ For no barrier to nucleation

$$2A\Delta G_V = \pi\sigma^2$$

$$2 \times 5 \times 10^{-10} \times \frac{1.7}{900}\Delta T \times 10^{+9} = \pi(0.2)^2 \qquad \text{Note } 1J = 1Nm$$

$$\Delta T = \frac{0.04\pi \times 900}{1.7} = \frac{0.04 \times 3.14159 \times 900}{1.7}$$

$$= \underline{\underline{66.5^\circ C}}$$

References

6.1 D. Turnbull, J.Chem Physics, 1952, vol 20, p.411.

6.2 R.B.Nicholson, "Phase Transformations", Metals Park 1968, ASM 1970.

6.3 A.J. Clemm and J.C. Fisher, Acta Met., 1955, Vol 3, pp 70-73.

7. The Zeldovitch Non-equilibrium Factor and Incubation Periods

7.1 The Steady State nucleation rate

Section 5.3 considered the metastable equilibrium concentration of embryos and gave the result:

$$C_n = N \exp\left(-\frac{\Delta G_n}{kT}\right) \qquad 5.6$$

In actual fact C_n will differ from this value near the critical nucleus size because some embryos will be dissolving while others will be growing. The actual concentration of embryos C'_n is illustrated in figure 7.1. The factor that allows for the equilibrium dissolution and growth of embryos up to critical nucleus size a* (containing n* atoms) is known as the Zeldovitch non-equilibrium factor Z.

The following treatment of this factor is taken from Aaronson and Lee's monograph (7.1).

If β_n = rate of growth of an embryo containing n atoms

α_{n+1} = rate of dissolution of an embryo containing (n+1) atoms

= rate at which atoms leave an embryo containing (n+1) atoms

C'_n = actual number of embryos /unit volume containing n atoms

C'_{n+1} = actual number of embryos /unit volume containing (n+1) atoms.

Then the net rate at which embryos containing n atoms become embryos containing (n+1) atoms is given by:-

$$J_n = \beta_n C'_n - \alpha_{n+1} C'_{n+1} \qquad 7.1$$

α_{n+1} is evaluated by using the principle of time reversal. This principle states that at equilibrium the forward and reverse reactions in a process occur exactly at the same rate

ie $J_n = 0 = \beta_n C_n - \alpha_{n+1} C_{n+1}$

$$\therefore \quad \alpha_{n+1} = \frac{\beta_n C_n}{C_{n+1}} \qquad 7.2$$

$\therefore$ substituting equation 7.2 in 7.1

$$J_n = \beta_n C'_n - \alpha_{n+1} C'_{n+1}$$

$$= \beta_n C'_n - \frac{\beta_n C_n}{C_{n+1}} C'_{n+1}$$

$$= \beta_n C_n \left\{\frac{C'_n}{C_n} - \frac{C'_{n+1}}{C_{n+1}}\right\} \qquad 7.3$$

It will be seen in figure 7.1 that for small embryos $\frac{C'_n}{C_n}$ is approximately constant except where the embryo size approaches $n^* - \delta/2$.

We can therefore express $\left\{\frac{C'_n}{C_n} - \frac{C'_{n+1}}{C_{n+1}}\right\}$ as a differential

ie $$\left\{\frac{C'_n}{C_n} - \frac{C'_{n+1}}{C_{n+1}}\right\} = \frac{\partial(C'_n/C_n)}{\partial n} \times \Delta n.$$

$$= \frac{\partial(C'_n/C_n)}{\partial n} \times [n - (n + 1)]$$

$$= - \frac{\partial(C'_n/C_n)}{\partial n} \quad \ldots\ldots\ldots\ldots\ldots\ldots\ldots\ldots\ldots \quad 7.4$$

Hence equation 7.3 becomes

$$J_n = \beta_n C_n \left\{\frac{C'_n}{C_n} - \frac{C'_{n+1}}{C_{n+1}}\right\}$$

$$= -\beta_n C_n \frac{\partial(C'_n/C_n)}{\partial n} \quad \ldots\ldots\ldots\ldots\ldots\ldots\ldots\ldots\ldots \quad 7.5$$

Now quasi-steady state equilibrium will occur when all values of C'_n are independent of time. This will result in J_n and the steady state nucleation rate J^*_n being independent of n and time.

Equation 7.4 therefore can be integrated with the following boundary conditions:-

$$\frac{C'_n}{C_n} \to 1 \qquad \text{as } n \to 1$$

and $$\frac{C'_n}{C_n} \to 0 \qquad \text{as } n \to \infty$$

Now β_n does not vary significantly with n and we can put it equal to β^*. Hence equation 7.5 can be re-arranged as

$$J_n \frac{dn}{C_n} = -\beta_n \, \partial(C'_n/C_n)$$

$$= -\beta^* \, \partial(C'_n/C_n) \quad \ldots\ldots\ldots\ldots\ldots\ldots\ldots\ldots\ldots \quad 7.6$$

Integrating equation 7.6 between the limits stated for quasi-steady state equilibrium:-

$$J^*\int_1^\infty \frac{dn}{C_n} = -\beta^*\int_1^0 d\left(\frac{C'_n}{C_n}\right)$$

$$= -\beta^*\left[\frac{C'_n}{C_n}\right]_1^0$$

$$= \beta^*$$

Hence $J^* = \beta^*/\int_1^\infty \frac{dn}{C_n}$ 7.7

Substituting equation 5.6, $C_n = N\exp(-\frac{\Delta G_n}{kT})$ in 7.7

$$J^* = N\beta^*/\int_1^\infty \exp(\frac{\Delta G_n}{kT})\,dn \qquad 7.8$$

In order to evaluate the integral, ΔG_n is expanded as a Taylor's series about n*, neglecting powers greater than the second

$$f(x) = f(x_o) + (x-x_o)f'(x_o) + \frac{(x-x_o)^2}{2} f''(x_o)$$

$$\therefore \quad \Delta G_n = \Delta G^* + \frac{(n-n^*)^2}{2}\frac{\partial^2(\Delta G_n)}{\partial n^2}\bigg|_{n^*} \qquad 7.9$$

Since $f(x_o) = \Delta G^*$ and $f'(x_o) = \frac{\partial(\Delta G_n)}{\partial n}\bigg|_{n^*} = 0$

at n = n* (because $\Delta G_n = \Delta G^*$, is a maximum at n*)

Hence,

$$\exp\left(\frac{\Delta G_n}{kT}\right) = \exp\left[\frac{\Delta G^*}{kT} + \frac{(n-n^*)^2}{2kT}\frac{\partial^2(\Delta G_n)}{\partial n^2}\bigg|_{n^*}\right]$$

$$= \exp\frac{\Delta G^*}{kT}\exp\left[\frac{(n-n^*)^2}{2kT}\,\partial^2\frac{(\Delta G_n)}{\partial n^2}\bigg|_{n^*}\right] \qquad 7.10$$

Now we put $\lambda^2 = (n-n^*)^2\pi Z^2$

where $Z^2 = \frac{-1}{2\pi kT}\,\partial^2\frac{(\Delta G_n)}{\partial n^2}\bigg|_{n^*}$ 7.11

ie $\quad \lambda = (n-n^*)Z\sqrt{\pi}$

$\therefore \quad d\lambda = Z\sqrt{\pi}\,dn$

when n = 1 $\quad \lambda = (1-n^*)Z\sqrt{\pi}$

$n = \infty \qquad \lambda = \infty$

Hence returning to the integral in equation 7.8

$$\int_1^\infty \exp\left(\frac{\Delta G_n}{kT}\right) = \int_1^\infty \exp \frac{\Delta G^*}{kT} \exp\left[\frac{(n-n^*)^2}{2kT} \partial^2 \frac{(\Delta G_n)}{\partial n^2}\bigg|_{n^*}\right] dn$$ - from equation 7.10

$$= \exp\left(\frac{\Delta G^*}{kT}\right)\int_1^\infty \exp \frac{(n-n^*)^2}{2kT} \partial^2 \frac{(\Delta G_n)}{\partial n^2}\bigg|_{n^*}\Bigg] dn$$

$$= \frac{1}{Z\sqrt{\pi}}\int_{(1-n^*).Z\sqrt{\pi}}^{\infty} \exp(-\lambda^2)\,d\lambda \quad \ldots\ldots\ldots\ldots \quad 7.12$$

substituting in the expressions for Z and λ.

Now the lower limit of the integral is in fact negative because $n^* > 1$, and because the integral is very close to zero for values of $\lambda < (1-n^*)Z\sqrt{\pi}$, the lower limit can be increased to $-\infty$ without introducing appreciable error.

$$\therefore \quad \int_1^\infty \exp\left[\frac{(n-n^*)^2}{2kT} \quad \frac{\partial^2(\Delta G_n)}{\partial n^2}\bigg|_{n^*}\right] dn$$

$$= \frac{1}{Z\sqrt{\pi}}\int_{(1-n^*)Z\sqrt{\pi}}^{\infty} \exp(-\lambda)^2\,d\lambda$$

$$= \frac{1}{Z\sqrt{\pi}}\int_{-\infty}^{\infty} \exp(-\lambda^2)\,d\lambda$$

$$= \frac{2}{Z\sqrt{\pi}}\int_{0}^{\infty} \exp(-\lambda^2)\,d\lambda \quad \ldots\ldots\ldots\ldots \quad 7.13$$

Now the integral $\frac{2}{\sqrt{\pi}}\int_0^\lambda \exp(-\lambda^2)\,d\lambda$ 7.14

= Error function, ϕ

The properties of this function are given in figure 7.2.

It will be seen from this figure that

$$\frac{2}{Z\sqrt{\pi}}\int_0^\lambda \exp(-\lambda^2)\,d\lambda = \frac{\phi}{Z} \quad \ldots\ldots\ldots\ldots \quad 7.15$$

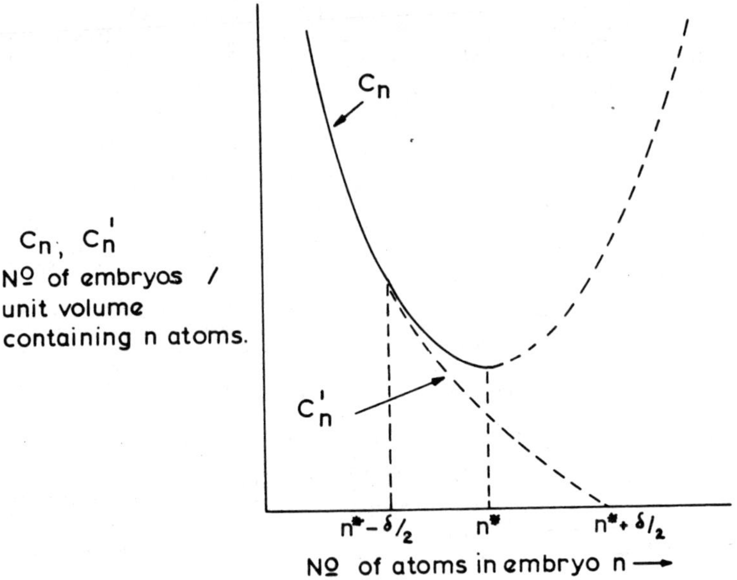

FIG. 7.1 Variation of concentration of embryos with size of embryos

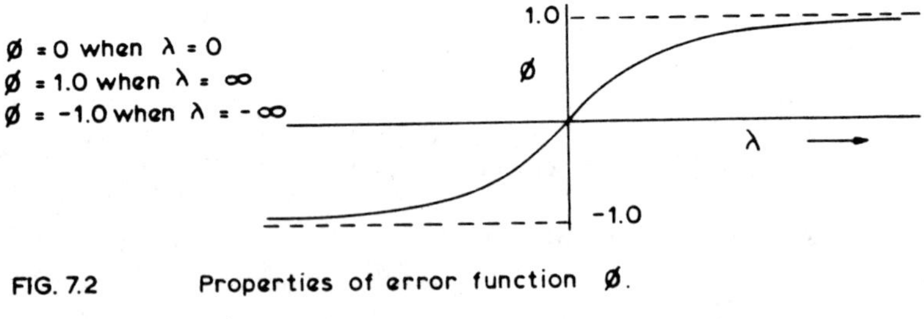

FIG. 7.2 Properties of error function ø.

$$\therefore \quad \frac{2}{Z\sqrt{\pi}} \int_1^{\infty} \exp(-\lambda^2)\,d\lambda = \frac{1}{Z}$$

$$\text{Hence} \int_1^{\infty} \exp\left[\frac{(n-n^*)^2}{2kT} \left.\frac{\partial^2(\Delta G_n)}{\partial n^2}\right|_{n^*}\right] dn = \frac{1}{Z}$$

∴ Substituting in the original integral, equation 7.12.

$$\int_1^{\infty} \exp\left(\frac{\Delta G_n}{kT}\right) = \exp\left(\frac{\Delta G^*}{kT}\right) \int_{-1}^{\infty} \exp\left[\frac{(n-n^*)^2}{2kT} \left.\frac{\partial^2(\Delta G_n)}{\partial n^2}\right|_{n^*}\right] dn$$

$$= \frac{1}{Z} \exp \frac{\Delta G^*}{kT}$$

Hence equation 7.5 becomes

$$J^* = N\beta^* \Big/ \int_1^{\infty} \exp\left(\frac{\Delta G_n}{kT}\right) dn$$

$$= ZN\beta^* \exp - \frac{\Delta G^*}{kT} \qquad \text{.......... 7.16}$$

7.2 Incubation Period

This arises from the time taken to achieve a steady state nucleation rate.

Physically it can be related to the time taken for an embryo to grow from one atom to the critical nucleus size n*.

Russell et al (7.2) adopted the following physical approach, figure 7.3.

They suggested that near the top of the activation energy barrier, embryos could be promoted to a size greater than critical by random walk across a distance δ defined in figure 7.3.

The thermal energy kT is sufficient to promote embryos of size $(n^* - \delta/2)$ to $(n^* + \delta/2)$.

The time required for this process is given by

$$\tau = \frac{\delta^2}{2\beta^*} \qquad \text{.......... 7.17}$$

Moreover Russell et al showed that τ is much treater than t' for most embryo ɕeometries.

This makes sense by considering the "principle of time reversal" which says that a process which occurs by statistical fluctuations takes place in

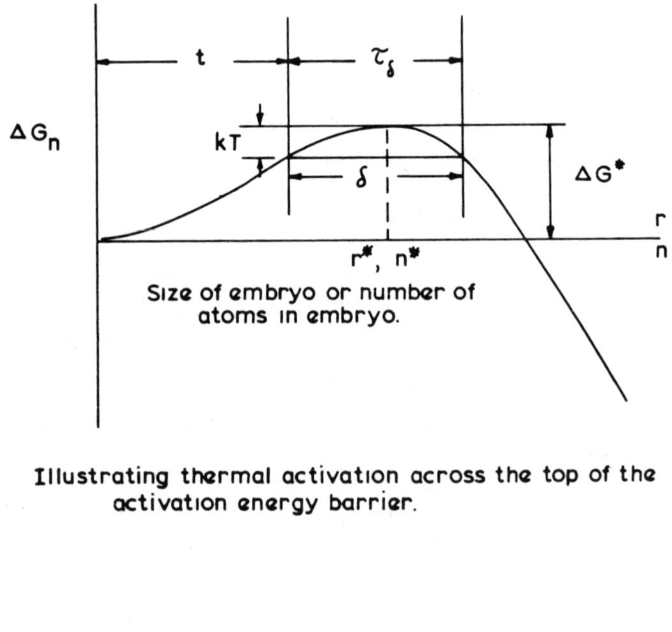

FIG. 7.3 Illustrating thermal activation across the top of the activation energy barrier.

the same time both forward and backward. Thus in the region t' ΔG_n is varying steeply with n and dissolution, and therefore growth of the embryos will occur fairly rapidly.

So it is a good approximation to take τ as the incubation period before steady state nucleation. This also overcomes the objection that we are considering growth of embryos from one atom. It is more likely that embryos occur due to dislocation interactions and these will be considerably larger than one atom.

δ can be evaluated as follows:-

Expanding ΔG_n in a Taylor series about n* we obtain as before, equation 7.9.

$$\Delta G_n = \Delta G^* + \frac{(n - n^*)^2}{2} \left(\frac{\partial^2 \Delta G_n}{\partial n^2}\right)_{n^*}$$

Putting $\Delta G^* = \Delta G + kT$

$n = n^* + \delta/2$ ie $n - n^* = \delta/2$

$$\therefore \quad \Delta G_n = \Delta G_n + kT + \frac{(\delta/2)^2}{2} \left(\frac{\partial^2 \Delta G_n}{\partial n^2}\right)_{n^*}$$

$$= \Delta G_n + kT + \frac{\delta^2}{8} \left(\frac{\partial^2 \Delta G_n}{\partial n^2}\right)_{n^*}$$

$$\therefore \quad \delta^2 = -8kT \left(\frac{\partial^2 \Delta G_n}{\partial n^2}\right)^{-1}_{n^*}$$

$$\text{ie} \quad \delta = \left[- \frac{1}{8kT} \left(\frac{\partial^2 \Delta G_n}{\partial n^2}\right)_{n^*}\right]^{-\frac{1}{2}} \quad \text{7.18}$$

Recall $Z^2 = \frac{-1}{2\pi kT} \left(\frac{\partial^2 \Delta G_n}{\partial n^2}\right)_{n^*}$

$$\text{ie} \quad Z = \left[\frac{-1}{6.3kT} \left(\frac{\partial^2 \Delta G_n}{\partial n^2}\right)_{n^*}\right]^{\frac{1}{2}}$$

$$\sim \frac{1}{\text{[illegible]}}$$

Some people then write for simplicity

$$\tau \sim (2\beta^* Z^2)^{-1} \quad \text{7.19}$$

<u>Problem 7.1</u> BSc II (Hons) Metallurgy and Microstructural Engineering, Physical Metallurgy, December 1979, Sheffield City Polytechnic.

Q.4 Calculate the critical size of a spherical nucleus of ferrite forming homogeneously from austenite at 860°C in pure iron. You may assume that the γ/α interface of the nucleus is incoherent and that there is negligible strain energy accompanying the formation of α.

Given that at 860°C, the Zeldovitch Factor Z for this nucleus = 2.7×10^{-7}, estimate the isothermal incubation period required for the formation of the nucleus at this temperature.

Comment on the size of the critical nucleus you obtain in your calculation.

DATA: Surface energy of incoherent γ/α interface = 0.56 J/m^2
Latent heat of transformation per unit volume of α (ΔH_v) = -1.27×10^8 J/m^3
Number of atoms/unit volume of γ (N_γ) = 8.7×10^{28}/m^3.

Frequency at which atoms attempt to join critical nucleus at 860°C = 7.64×10^7 s^{-1}.

Free energy of embryo radius r

$$\Delta G = \frac{4}{3}\pi r^3 \Delta G_v + 4\pi r^2 \sigma$$

At critical nucleus size r*, $\frac{\partial \Delta G}{\partial r} = 0$.

$$\therefore \quad \frac{\partial \Delta G}{\partial r} = 4\pi r^2 \Delta G_v + 8\pi r\sigma = 0 \text{ at } r^*$$

$$\therefore \quad r^* = \frac{-2\sigma}{\Delta G_v}$$

Now $\quad \Delta G_v = \Delta H_v - T\Delta S_v = 0$ at T_o

$$\therefore \quad \Delta S_v = \frac{\Delta H_v}{T_o}$$

and

$$\Delta G_v = \Delta H_v - T\Delta S_v = \Delta H_v - \frac{T\Delta S_v}{T_o}$$

$$= \frac{\Delta H_v \Delta T}{T_o}$$

$$= \frac{-1.27 \times 10^8 \times 500}{1183}$$

$T_o = 910 + 273$
$= 1183$ K

$$\therefore \quad r^* = \frac{2 \times 0.56 \times 1183}{1.27 \times 10^8 \times 50}$$

$$= 2.0865 \times 10^{-7}\text{m}$$

$$\simeq \underline{\underline{2,100 \text{ Å}}}$$

Rate of growth of critical nucleus β^* = No. of atoms on surface of nucleus of critical size x frequency at which atoms attempt to join critical nucleus

$= 4\pi r^2 (N_\gamma)^{2/3} \times \nu$

$= 4\pi(2.0865 \times 10^{-7})^2 \times (8.7 \times 10^{28})^{2/3} \times 7.64 \times 10^7$

$= 8.21 \times 10^{14}$

$\therefore$ Incubation period $\tau \simeq (2\beta^* Z^2)^{-1}$

$= [2 \times 8.21 \times 10^{14} \times (2.7 \times 10^{-7})^2]^{-1}$

$= (119.64)^{-1}$ seconds

= 8.4 milliseconds

The critical nucleus size is far too large for what can be expected to be observed in practice. Nucleation in practice will occur hetrogeneously at grain corners and grain edges which will reduce the size of the critical nucleus.

References

7.1 H.I. Aaronson and J.K. Lee, "Lectures on the Theory of Phase Transformations", Edited by H.I. Aaronson, AIME, 1975.

7.2 K.C. Russell, "Phase Transformations", Metals Park. 1968, ASM, 1970.

8. Spinodal Decomposition

8.1 Definition of Spinodal

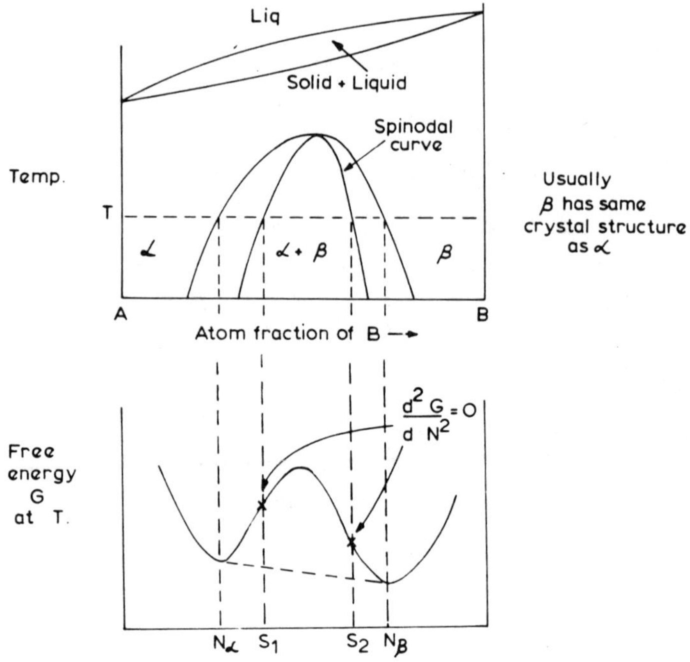

FIG. 8.1 Definition of spinodal

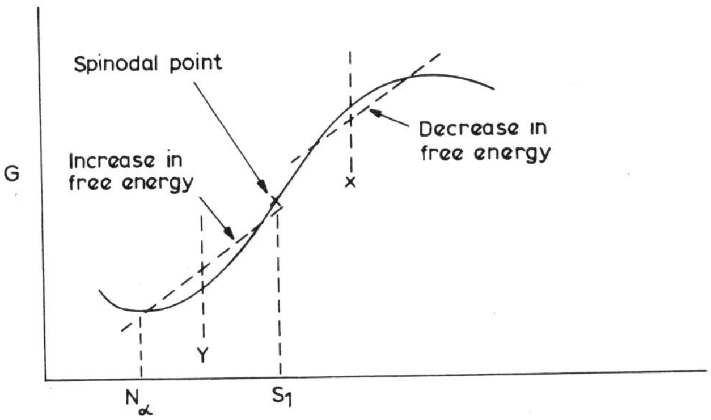

FIG. 8.2 Enlargement of free energycurve in vicinity of spinodal showing fluctuations in compositions within the spinodal lead to a decrease in free energy x, whereas fluctuations outside the spinodal Y, lead to an increase in free energy and an energy barrier to nucleation.

Since there is no thermodynamic barrier inside the spinodal, the decomposition is determined solely by diffusion. We therefore have to solve the relevant diffusion equation to determine how spinodal decomposition occurs.

8.2 Thermodynamic diffusion equation

The classical diffusion equations, Fick's 1st and 2nd Laws are written in terms of concentration. However the rate of diffusion in metals depends upon the activity gradient or gradient in chemical potential rather than concentration gradients. This explains the Kirhendal effect and Darken's experiment.

Consider an alloy consisting of components 1 and 2.

N_1 = atom fraction of component (1)

N_2 = $(1-N_1)$ = atom fraction of component (2)

N_v = No. of atoms/unit volume.

μ_1 = chemical potential of component (1) = dG_1/dN_1

Flux of atoms (1)

$$J_1 = -N_v\, N_1\, v_1 \frac{\partial \mu_1}{\partial x}$$

where v_1 = mobility of component 1 for unit chemical potential gradient/ atom of (1).

The term N_v arises because units of chemical potential are energy per atom and we require per unit volume.

$$J_1 = -N_v\, N_1\, v_1 \frac{\partial \mu_1}{\partial N_1}\frac{\partial N_1}{\partial x}$$

cf $$J = -D\frac{\partial c}{\partial x}$$

$$\therefore \quad D_1 = N_v\, N_1\, v_1 \frac{\partial \mu_1}{\partial N_1} = N_1\, N_2\, v_1 \frac{d^2 G_v}{dN_1^{\,2}}$$

G_v = free energy/unit volume

Now free energy/atom $G = N_1\mu_1 + N_2\mu_2$

$$\therefore \quad \text{Free energy/unit volume } G_v = N_v\left[N_1\mu_1 + N_2\mu_2\right]$$

$$\therefore \quad \frac{dG_v}{dN_1} = N_v\{\mu_1 + N_1 d\mu_1 + d\mu_2 - N_1 d\mu_2 - \mu_2\}$$

Gibb's Duhem equation gives

$$N_1 d\mu_1 + N_2 d\mu_2 = 0$$

$$\therefore \quad \frac{dG_v}{dN_1} = N_v\{\mu_1 - \mu_2\}$$

$$\therefore \quad \frac{d^2G_V}{dN_1^2} = N_v\ (d\mu_1 - d\mu_2)$$

$$N_2 \frac{d^2G_V}{dN_1{}^2} = N_v\ (N_2 d\mu_1 - N_2 d\mu_2)$$

$$= N_v\ \{(1-N_1)d\mu_1 - N_2 d\mu_2\}$$

$$= N_v\ \{d\mu_1 - N_1 d\mu_1 - N_2 d\mu_2\}$$

$$= N_v d\mu_1 \quad \text{(from Gibb's Duhem)}$$

$$\therefore \quad D_1 = N_v\ N_1\ v_1 \frac{d\mu_1}{dN_1} = N_1\ N_2\ v_1 \frac{d^2G_V}{dN_1{}^2}$$

It can be shown that there is a mean diffusion coefficient $\bar{D}$ given by

$$\bar{D} = \frac{N_2 D_1}{N_v} + \frac{N_1 D_2}{N_v}$$

N_v arises because N_1 & N_2 are atom fractions and D is required in dimensions of distance

ie $\frac{N_1}{N_v}$ = volume occupied by N_1 atoms of (1)

$\bar{D}$ is related to concentration by

$$\frac{dN_1}{dt} = \bar{D} \frac{\partial^2 N_1}{\partial x^2}$$

$$\bar{D} = \frac{N_2}{N_v} D_1 + \frac{N_1}{N_v} D_2 = \frac{N_1 N_2}{N_v} \frac{d^2G_V}{dN^2} \{N_2\ v_1 + N_1\ v_2\}$$

Hence solution of this equation $\frac{dN}{dt} = \bar{D} \frac{d^2N}{dx^2}$ is of the form

$N = f(x, t)$ with a periodic variation in composition with distance and time, provided $\frac{d^2G_V}{dN^2} < 0$ ie -ve - within spinodal.

For this equation the wavelength λ of the fluctuations $\sim$ lattice parameter.

In practice the wavelength of the fluctuations $\sim$50 - 200Å.

This is because of two effects which oppose the change in free energy.

1) The variation in composition with distance can be regarded as producing a diffuse interface with a surface energy opposing transformation.

2) The variation in composition will also produce variations in lattice parameter and so change volume/atom and produce coherency strains. These strains will also oppose the chemical driving force for transformation.

A high chemical driving force gives rise to short wavelengths λ_m.

Consequently we can say

λ_m decreases with (a) decreasing temperature

(b) as the average composition approaches the centre of the spinodal.

All these factors increase the chemical driving force.

Note in practice the modulations have particular orientations, this is because the coherency strains are crystallographic, which in turn is due to the variation of elastic modulii with orientation.

The spinodal decomposition obeys a typical 'C' curve shape, however with very short incubation periods $\sim$ 1/10th of a second.

Just below the spinodal temp, there is only a small driving force, therefore the reaction is slow and the wavelength of the modulations large.

At large degrees of undercooling the driving force is large but diffusion is sluggish and therefore the reaction slower and wavelength smaller.

References

8.1 "Phase Transformations", Metals Park, 1968, ASM 1970.

9. Massive Transformations in Iron

This is a rapid transformation in which there is no change in composition. At high degrees of cooling the massive transformation can be suppressed and a martensitic transformation obtained at a lower temperature (9.1, 9.2, 9.3, 9.4 and 9.5).

The transformation is thought to occur by nucleation and rapid movement of an incoherent boundary (>0.44 mm/s in Fe-0.01%C (9.6)) with atom transfer across the boundary and no long range diffusion (9.3). Nucleation is therefore thought to be the rate controlling factor in this type of transformation (9.7, 9.8).

The transformation was first discovered by Philips (9.7) in β brass and subsequently in Al-Cu alloys by Greninger (9.1) who was the first author to use the term "massive transformation" for this type of structure. Massalski (9.10) showed in Cu-Ga alloys that the products of the massive transformation cross the parent grain boundaries. This was subsequently shown to be the case for the massive transformation in Fe-Ni alloys (9.11, 9.12,). The most recent review of massive transformations is given in 9.3.

Gilbert and Owen (9.4) showed in Fe-Ni and Fe-Cr alloys that on plotting transformation temperature as a function of cooling rate that the transformation is suppressed to a constant value or plateau. By comparison of the plateaux temperatures obtained in Fe-Cr alloys (9.4) with the TTT curve for an Fe-8.5%Cr - 0.05%C alloy (9.13), Wilson concluded that such temperature plateaux correspond to the nose of the 'C' curve on the TTT diagram and hence the maximum rate of transformation. This viewpoint is disputed by some authors (9.14).

Parr and his co-workers obtained two pleateaux in their study of pure iron (9.5) and Fe-Ni alloys (9.15) at a lower temperature than that of Gilbert and Owen. Bibby and Parr thought that their second pleateau corresponded to a martensitic transformation and that as the carbon content becomes less than 0.005%C M_s in pure iron rises to $\sim 700^{\circ}C$. However Wilson observed four temperature plateaux in his work on iron containing 0.01%C, figure 9.1 (9.8) and tried to rationalise his results and those of Bibby and Parr (5) as follows:

1) $860^{\circ}C$ Equiaxed ferrite - a massive transformation occuring by nucleation of an incoherent grain boundary allotromorph, figure 9.2, and rapid movement of the incoherent boundary to give an equiaxed grain structure.

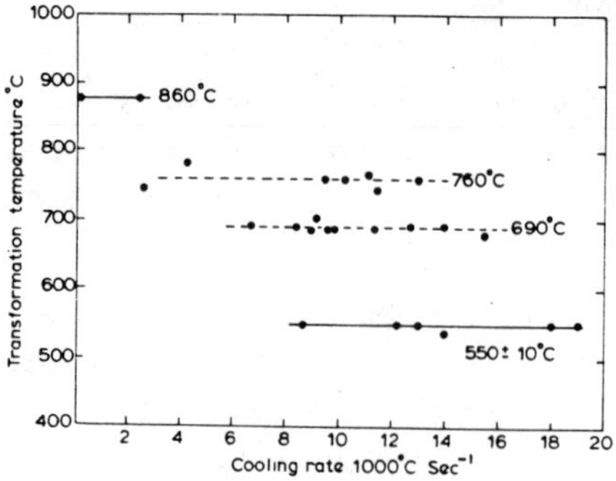

Fig. 9.1(a) Variation of transformation temperature with cooling rate for BISRA Fe ARL8.

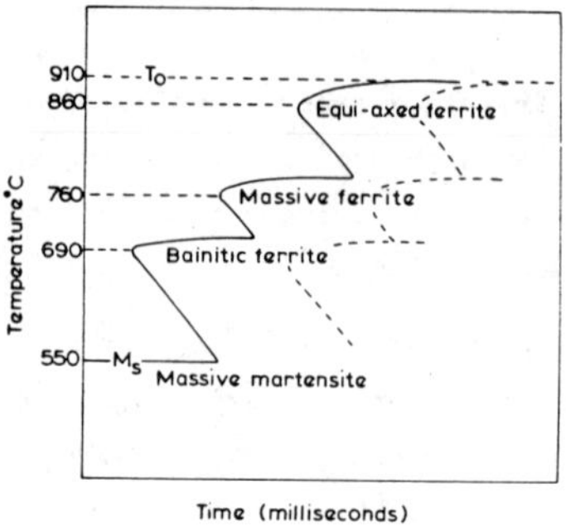

Fig. 9.1(b) Schematic TTT diagram for pure iron.

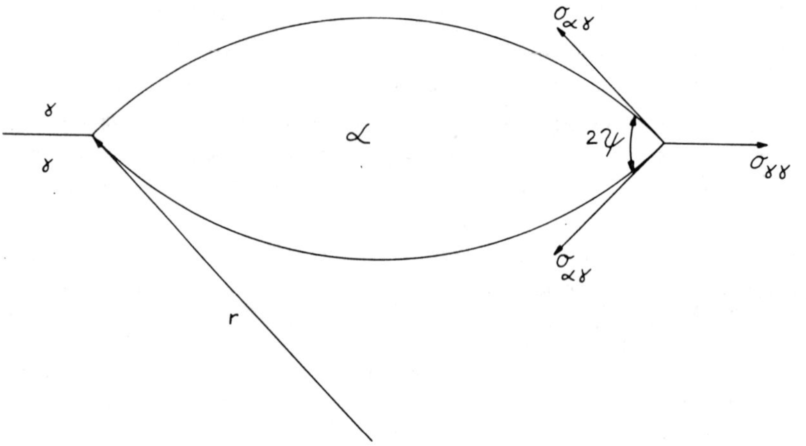

FIG. 9.2 PROPOSED NUCLEUS FOR THE EQUI-AXED FERRITE TRANSFORMATION

2) 760°C Massive ferrite - another massive transformation occurring by nucleation of a coherent grain boundary allotromorph, figure 9.3, and rapid movement of the incoherent surface to give small grains of ragged outline.
3) 690°C A banitic transformation - an accicular structure which is thought to occur by movement of ledges giving rise to surface tilts on a pre-polished surface and hence could be mistaken for a martensitic transformation.
4) 550°C Lath or Massive martensite.
5) 420°C Twinned Martensite - the following authors have presented evidence that the lath martensitic transformation can be suppressed and twinned martensite obtained at a lower temperature (9.16, 9.17, 9.8, 9.18).

The incubation periods of the 'C' curve are thought to be very sensitive to carbon content and thus in some irons only one or two temperature plateaux are obtained.

Thus the micrograph of surface tilts in an Fe-15%Ni transformed at 388°C with a cooling rate of $\sim$100°C/sec in reference (9.19) is now thought to be a bainitic transformation and not lath or massive martensite, lath martensite being obtained at a higher cooling rate and lower temperature.

Problem 9.1

Calculate the rate of growth of the massive transformation in pure iron at 860°C from the following data:

Pre-exponential diffusion coefficient for grain boundary diffusion
$D_o = 3.4\ cm^2/s$.
Lattice parameter of austenite $a_\gamma = 3.64 \times 10^{-8}$ cms
Lattice parameter of ferrite $a_\alpha = 2.90 \times 10^{-8}$ cms
Chemical driving force at 860°C = 9.75 cals/mol
= 6.78×10^{-23} J/atom
Activation energy for grain boundary diffusion in austenite
$Q = 2.71 \times 10^{-19}$ J/atom

The following derivation is due to Burke (9.20).
The number of atoms leaving per unit area

$$= p_\gamma A_\gamma N_\gamma \nu_\gamma \exp - \frac{G^{\gamma\to\alpha}}{kT} \quad \ldots\ldots\ldots\ldots \quad 9.1$$

where p_γ ($\sim\frac{1}{6}$) is the probability that a vibration is in the right direction.

ν_γ is the frequency of vibration of the atoms

N_γ is the number of atoms/unit area of γ at the interface.

A_γ is the accommodation coefficient for the γ crystal, ie the fraction of the numbers of sites on the surface at which atoms can be accommodated into the growing lattice.

Similarly the number of atoms leaving α per unit area

$$= p_\alpha A_\alpha N_\alpha \nu_\alpha \exp - \frac{G^{\alpha\to\gamma}}{kT} \quad \ldots\ldots\ldots\ldots \quad 9.2$$

Hence the net reaction in the direction

$$= p_\alpha A_\alpha N_\alpha \nu_\alpha \exp - \frac{G^{\gamma\to\alpha}}{kT} - p_\gamma A_\gamma N_\gamma \nu_\gamma \exp - \frac{G^{\alpha\to\gamma}}{kT}$$

$$= pAN\nu \{ \exp - \frac{G^{\gamma\to\alpha}}{kT} - \exp - \frac{G^{\alpha\to\gamma}}{kT} \}$$

$$= pAN\nu \{ \exp - \frac{G^{\gamma\to\alpha}}{kT} - \exp - \frac{(G^{\gamma\to\alpha} + \Delta G)}{kT} \} \quad \ldots\ldots\ldots\ldots \quad 9.3$$

Putting $p_\alpha = p_\gamma = p$; $A_\alpha = A_\gamma = A$; $N_\alpha = N_\gamma = N$

$\nu_\alpha = \nu_\gamma = \nu$

which is a reasonable approximation for polymorphic transformations.

Equation 9.3 gives the number of atoms gained by α per unit area per unit time. Hence if we multiply this by the volume of one atom in α, V we obtain the rate of advance of the interface. The equation simplifies to

$$\dot{R} = \delta\nu \{ 1 - \exp - \frac{\Delta G_v}{kT} \} \exp - \frac{Q}{kT} \quad \ldots\ldots\ldots\ldots \quad 9.4$$

where δ = thickness of γ/α interface

Q = activation energy for atom transfer across α/γ interface

$\equiv$ activation energy for grain boundary diffusion in austenite

ν = frequency of vibration of atoms

$$= \frac{D_o}{(a_\gamma)^2} = \frac{3.4}{(3.64 \times 10^{-8})^2}$$

$$= 2.57 \times 10^{15}\ s^{-1}$$

$$\delta = \frac{(3.64 + 2.9)}{2} \times 10^{-8}\ cm = 3.27 \times 10^{-8}\ cm$$

$$\therefore \quad \dot{R} = \delta\nu \{ 1 - \exp - \frac{\Delta G_v}{kT} \} \exp - \frac{Q}{kT}$$

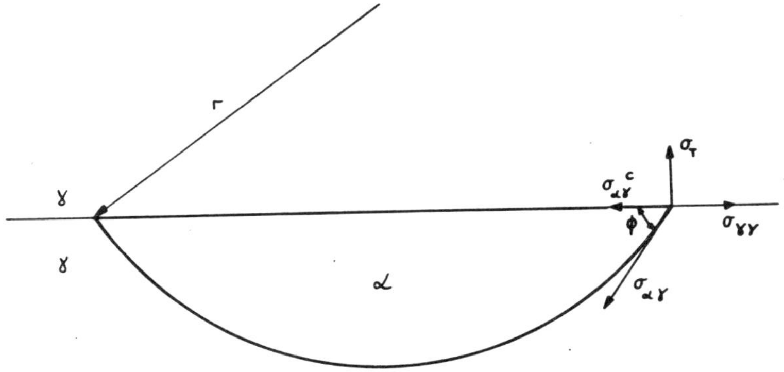

FIG. 9.3 PROPOSED NUCLEUS FOR THE MASSIVE FERRITE TRANSFORMATION.

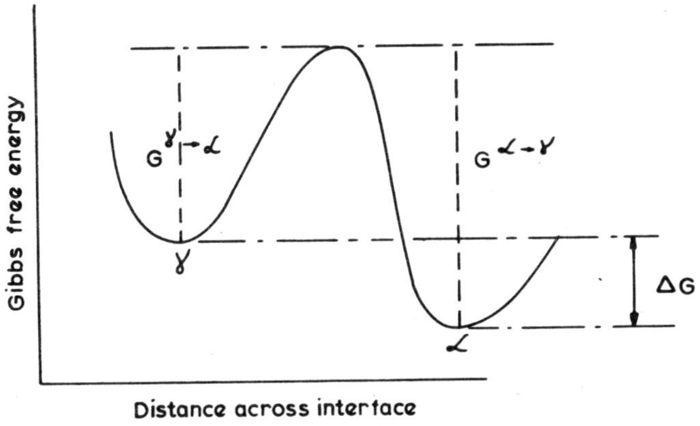

FIG. 9.4 Atom transfer across the α / γ interface during the massive transformation.

$$= 3.27 \times 10^{-8} \times 2.57 \times 10^{15} \{ 1 - \exp - \frac{6.78 \times 10^{-23}}{1.38042 \times 10^{-23} \times 1133} \} \exp - \frac{2.71 \times 10^{-19}}{1.38042 \times 10^{-23} \times 1133}$$

$$= 1.085 \times 10^{-2} \text{ cm/sec}$$

$$= \underline{\underline{0.11 \text{ mm/sec}}}$$

This may be compared with the experimental value of 0.44 mm/s (9.6).

Nucleation of equi-axed ferrite

Bhattacharyya et al (9.21) suggested that because of the rapid growth of the massive transformation, nucleation was the critical event in the formation of phases by the massive transformation. They calculated the incubation period for growth of the nucleus given in figure 9.2 from zero to the critical nucleus size. However their analysis did not predict a transformation plateau, since they assumed a constant value, independent of temperature for the frequency ν at which atoms joined the embryo. The author (9.22, 9.8) used Russell's treatment of incubation periods (9.23) and allowing for the temperature dependance of ν was able to predict a plateau at 789°C.

Problem 9.2

Determine the plateau temperature for equi-axed ferrite as a function of cooling rate in pure iron.

Data $T_o = 910^oC = 1183K$; $\Delta H_v = -1.27 \times 10^8$ J/m^3

$\sigma_{\gamma\alpha} = 0.85$ J/m^2; $\sigma_{\alpha\gamma} = 0.56$ J/m^2

$a_\alpha = 3.64 \times 10^{-10}$m; $a_\gamma = 2.9 \times 10^{-10}$m

For grain boundary diffusion on austenite

$Q = 2.71 \times 10^{-19}$ J/atom; $D_o = 3.4$ cm^2/s

The model for the nucleus is given in figure 9.2 and the volume and surface areas for half this nucleus is given in figure 9.5.

The incubation period τ for the formation of the nucleus is given by equations 7.12 and 7.17,

$$\text{ie} \quad \tau = \frac{\delta^2}{2\beta^*}; \quad \delta = \left[- \frac{1}{8kT} \left(\frac{\partial^2 \Delta G_n}{\partial n^2} \right)_{n^*} \right]^{-\frac{1}{2}}$$

The free energy accompanying formation of the embryo shapes in figure 9.2 is as follows:

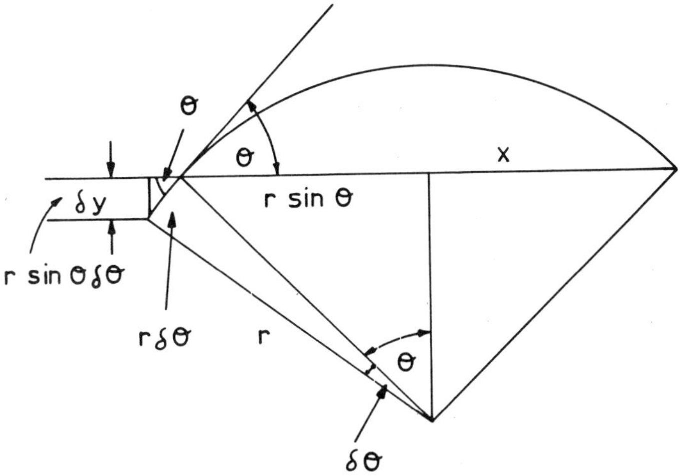

FIG. 9.5 Volume and surface area of semi-spherical cap.

Surface area of increment $= 2\pi x r\delta\theta$

$= 2\pi r\sin\theta r\delta\theta$

$\therefore$ Total surface area $= \int_0^\theta 2\pi r^2\sin\theta d\theta$

$= \left[-2\pi r^2\cos\theta\right]_0^\theta = \underline{\underline{2\pi r^2(1-\cos\theta)}}$

Volume of increment $= \pi x^2\delta y$

$= \pi r^2\sin^2\theta r\sin\theta\delta\theta$

$\therefore$ Total volume $= \int_0^\theta \pi r^3\sin^3 d\theta$

$= \pi r^3\int_0^\theta \sin\theta(1-\cos^2\theta)d\theta$

$= \pi r^3\int_0^\theta \sin\theta d\theta - \pi r^3\int_0^\theta \sin\theta\cos^2\theta d\theta$

Put $n = \cos\theta$ then $dn = -\sin\theta d\theta$

Total volume $= \pi r^3\left[-\cos\theta\right]_0^\theta + \int_{\theta=0}^\theta n^2 dn$

$= \pi r^3(1-\cos\theta) + \left[\frac{n^3}{3}\right]_{\theta=0}^\theta$

$= \pi r^3(1-\cos\theta) + \left[\frac{\cos^3\theta}{3}\right]_0^\theta$

$= \pi r^3\{1-\cos\theta + \frac{\cos^3\theta}{3} - 1/3\} = 4/3\pi r^3\frac{(2-3\cos\theta + \cos^3\theta)}{4} = \underline{\underline{4/3\pi r^3 f(\theta)}}$

$$\Delta G = 8/3\pi r^3 f(\psi)\Delta G_v + 4\pi r^2(1 - \cos\psi)\sigma_{\alpha\gamma} - \pi r^2 \sin^2\psi\sigma_{\gamma\gamma}$$

It may be assumed that the volume change accompanying formation of the embryo is accommodated by the two incoherent interfaces, and therefore the strain energy W is zero.

Further resolving the surface energies in figure 9.2

$$\sigma_{\gamma\gamma} = 2\sigma_{\alpha\gamma}\cos\psi \qquad \therefore \cos\psi = \frac{\sigma_{\gamma\gamma}}{2\sigma_{\alpha\gamma}} = 0.759$$

$$\therefore \quad \Delta G = 8/3\pi r^3 f(\psi)\Delta G_v + 4\pi r^2(1 - \cos\psi)\sigma_{\alpha\gamma} - \pi r^2(1 - \cos^2\psi)2\sigma_{\alpha\gamma}\cos\psi$$

$$= 8/3\pi r^3 f(\psi)\Delta G_v + 2\pi r^2\sigma_{\alpha\gamma}\{2 - 2\cos\psi - \cos\psi + \cos^3\psi\}$$

$$= 8/3\pi r^3 f(\psi)\Delta G_v + 2\pi r^2\sigma_{\alpha\gamma}\{2 - 3\cos\psi + \cos^3\psi\}$$

$$= 8/3\pi r^3 f(\psi)\Delta G_v + 8\pi r^2\sigma_{\alpha\gamma} f(\psi) \qquad \text{.......................... 9.5}$$

Note $f(\psi) = \dfrac{2 - 3\cos\psi + \cos^3\psi}{4} = 0.04$

If N_α = No of atoms per unit volume of α

= (1 + 8 x 1/8) atoms/unit cell

$$= \frac{2}{(2.9 \times 10^{-10})^3} = 8.2 \times 10^{28} \text{ atoms/m}^3$$

Then No of atoms in embryo $n = 8/3\pi r^3 f(\psi) N_\alpha$

$$\text{ie} \quad r^3 = \frac{3}{8\pi}\frac{n}{f(\psi)N_\alpha} \qquad \text{.. 9.6}$$

We can express ΔG in terms of n

$$\Delta G_n = \frac{n}{N_\alpha}\Delta G_v + 8\pi\sigma_{\alpha\gamma} f(\psi)\left(\frac{3}{8\pi}\frac{n}{f(\psi)N_\alpha}\right)^{2/3}$$

$$= \frac{n}{N_\alpha}\Delta G_v + (8\pi f(\psi))^{1/3}\sigma_{\alpha\gamma}\left(\frac{3n}{N}\right)^{2/3} \qquad \text{...................... 9.7}$$

$$\frac{\partial \Delta G_n}{\partial n} = \frac{\Delta G_v}{N_\alpha} + 2/3(8\pi f(\psi))^{1/3}\sigma_{\alpha\gamma}\left(\frac{3}{N_\alpha}\right)^{2/3} n^{-1/3}$$

$$= \frac{\Delta G_v}{N_\alpha} + \left\{\frac{64\pi f(\psi)}{3}\right\}^{1/3}\frac{\sigma_{\alpha\gamma}}{N_\alpha^{2/3}} n^{-1/3} \qquad \text{...................... 9.8}$$

$$\frac{\partial^2 \Delta G_n}{\partial n^2} = -1/3\left\{\frac{64\pi f(\psi)}{3}\right\}^{1/3}\frac{\sigma_{\alpha\gamma}}{N_\alpha^{2/3}} n^{-4/3} \qquad \text{...................... 9.9}$$

$$\text{At } n^* \ \frac{\partial \Delta G}{\partial n} = 0 \qquad \therefore \frac{\Delta G_n}{N_\alpha} = \left\{\frac{64\pi f(\psi)}{3}\right\}^{1/3}\frac{\sigma_{\alpha\gamma}}{N_\alpha^{2/3}} n^{*-1/3}$$

$$\therefore \quad n^{*1/3} = \frac{N_\alpha}{\Delta C_v} \left\{\frac{64\pi f(\psi)}{3}\right\}^{1/3} \frac{\sigma_{\alpha\gamma}}{N_\alpha^{2/3}}$$

$$= \left\{\frac{64\pi f(\psi)}{3}\right\}^{1/3} N_\alpha^{1/3} \frac{\sigma_{\alpha\gamma}}{\Delta G_v}$$

$$\therefore \quad n^* = \frac{64\pi f(\psi)}{3} N_\alpha \left(\frac{\sigma_{\alpha\gamma}}{\Delta G_v}\right)^3 \quad \ldots\ldots\ldots\ldots \quad 9.10$$

Substituting for n* in equation 9.9

$$\left(\frac{\partial^2 \Delta G_n}{\partial n^2}\right)_{n^*} = -1/3\left\{\frac{64\pi f(\psi)}{3}\right\}^{1/3} \frac{\sigma_{\alpha\gamma}}{N_\alpha^{2/3}} \left\{\frac{64\pi f(\psi)}{3}\right\}^{-4/3} N_\alpha^{-4/3}\left(\frac{\sigma_{\alpha\gamma}}{\Delta G_v}\right)^{-4}$$

$$= -1/3\left\{\frac{64\pi f(\psi)}{3}\right\}^{-1} N_\alpha^{-2/3} N_\alpha^{-4/3} \frac{\Delta G_v^4}{\sigma_{\alpha\gamma}^3}$$

$$= -\frac{1}{64\pi f(\psi)} \frac{1}{N_\alpha^2} \frac{\Delta G_v^4}{\sigma_{\alpha\gamma}^3}$$

$$\therefore \quad \delta = \left[-\frac{1}{8kT}\left(\frac{\partial^2 \Delta G_n}{\partial n^2}\right)_{n^*}\right]^{-1/2}$$

$$= \left[+\frac{1}{8kT}\frac{\Delta G_v^4}{64\pi f(\psi) N_\alpha^2 \sigma_{\alpha\gamma}^3}\right]^{-1/2} \quad \ldots\ldots\ldots\ldots \quad 9.11$$

$$\therefore \quad \delta^2 = +\frac{8kT 64\pi f(\psi) N_\alpha^2 \sigma_{\alpha\gamma}^3}{\Delta G_v^4} \quad \ldots\ldots\ldots\ldots \quad 9.12$$

β* = No. of atoms on surface of nucleus of critical size x rate at which atoms join nucleus

$$= 4\pi r^{*2}(1 - \cos\theta) N_\gamma^{2/3} \frac{D_o}{a_\gamma^2} \exp - \frac{Q}{kT}$$

$$= 4\pi\left(\frac{3}{8\pi}\frac{n^* N_\gamma}{f(\psi) N_\alpha}\right)^{2/3} (1 - \cos\psi) \frac{D_o}{a_\gamma^2} \exp - \frac{Q}{kT}$$

from equation 9.6.

Substituting ror n*, equation 9.10

$$\beta^* = 4\pi\left\{\frac{3}{8\pi}\frac{N_\gamma}{f(\psi) N_\alpha} \times \frac{64\pi f(\psi) N_\alpha}{3}\left(\frac{\sigma_{\alpha\gamma}}{\Delta G_v}\right)^3\right\}^{2/3} (1 - \cos\psi) \frac{D_o}{a_\gamma^2} \exp - \frac{Q}{kT}$$

$$= 4\pi\left(\frac{2\sigma_{\alpha\gamma}}{\Delta G_v}\right)^2 (1 - \cos\psi) N_\gamma^{2/3} \frac{D_o}{a_\gamma^2} \exp - \frac{Q}{kT} \qquad 9.13$$

Substituting 9.12 and 9.13 in $\tau = \frac{\delta^2}{2\beta^*}$, yields

$$\tau = 8kT \frac{64\pi f(\psi) N_\alpha^2 \sigma_{\alpha\gamma}^3}{\Delta G_v^4} \cdot \frac{1}{8\pi}\left(\frac{\Delta G_v}{2\sigma_{\alpha\gamma}}\right)^2 N_\gamma^{2/3} \frac{1}{(1 - \cos\psi)} \frac{a_\gamma^2}{D_o} \exp \frac{Q}{kT}$$

$$= 16 \frac{N_\alpha^2}{N_\gamma^{2/3}} kT \frac{f(\psi)}{(1 - \cos\psi)} \sigma_{\alpha\gamma} \Delta G_v^2 \frac{a_\gamma^2}{D_o} \exp \frac{Q}{kT} \qquad 9.14$$

Substituting $\Delta G_v = \frac{\Delta H_v \Delta T}{T_o}$ given

$$\tau = B \frac{T}{\Delta T^2} \exp \frac{Q}{kT} \qquad 9.15$$

where $B = 16 \frac{N_\alpha^2}{N_\gamma^{2/3}} k \frac{f(\psi)}{(1 - \cos\psi)} \sigma_{\alpha\gamma} \frac{T_o^2}{\Delta H_v} \frac{a_\gamma^2}{D_o}$

We now have to find the minimum value of τ as a function of degree of undercooling ΔT, hence differentiating equation 9.15 with respect to temperature

$$\frac{d\tau}{dT} = \frac{1}{\Delta T^2} \exp \frac{Q}{kT} + \frac{2T}{\Delta T^3} B \exp \frac{Q}{kT} + \frac{BT}{\Delta T^2} \left\{- \frac{Q}{kT^2} \exp \frac{Q}{kT}\right\}$$

$$= \frac{B}{\Delta T^2} \exp \frac{Q}{kT} \left\{1 + \frac{2T}{\Delta T} - \frac{Q}{kT}\right\}$$

Hence at the minimum value of τ

$$\left\{1 + \frac{2T}{\Delta T} - \frac{Q}{kT}\right\} = 0.$$

Putting $\Delta T = T_o - T$ and multiplying through by $T(T_o - T)$ yields

$$T(T_o - T) + 2T^2 - \frac{Q}{k}(T_o - T) = 0$$

$$T T_o - T^2 + 2T^2 - \frac{Q}{k} T_o + \frac{Q}{k} T = 0$$

ie $\quad T^2 + T\left(T_o + \frac{Q}{k}\right) - \frac{Q}{k} T_o = 0$

cf $\quad ax^2 + bx + c = 0$

$\therefore$ Plateau temperature $T_m = \frac{1}{2}\left[-\left(T_o + \frac{Q}{k}\right) \pm \sqrt{T_o^2 + \frac{Q^2}{k^2} + \frac{6QT_o}{k}}\right]$

$= 1062$ K

$= \underline{\underline{789^\circ C}}$

This may be compared with the value of 800°C reported by the Russian workers for the equi-axed ferrite transformation in zone refined iron (9.24). However a criticism of this model is that it yields large critical nuclei sizes ∿1,290Å. Perhaps a more reasonable nucleus size may be obtained from grain corner nuclei.

Nucleation of massive ferrite

Wilson (9.22) suggested that figure 9.3 was a suitable model for the nucleus of this transformation. In this case the strain energy W due to the volume change cannot be neglected and can be computed from the volume change at the plateau temperature. The semi-coherent boundary is thought not to move or only slightly during growth of this nucleus, the major growth occurring by movement of the semi-spherical incoherent boundary. A similar analysis to that outlined in problem 9.2, but allowing for strain energy W yields a value of 748°C for the plateau temperature. However once again the critical nucleus size (∿250Å) is still rather large.

References

9.1 A.B. Greninger, Trans AIME, 1939, Vol 133, p 204.

9.2 D. Hull and R.D. Garwood, "The Mechanism of Phase Transformations in Metals", Institute of Metals, London, 1956, p 219.

9.3 T.B. Massalski, "Phase Transformations", ASM 1970, p 437.

9.4 A. Gilbert and W.S. Owen, Acta Met, 1962, vol 10, p 45.

9.5 M.J. Bibby and J. Gordon Parr. JISI, 1964, vol 202, p 100.

9.6 E.O. Rasanen, Doctor of Technology Thesis 1969, Technical University, Otaniemi, Helsinki.

9.7 S.K. Bhattacharya, J.H. Perepuzko and T.B. Massalshi, Acta. Met., 1974, Vol 22, p 879.

9.8 E.A. Wilson, S.M. Vickers, C. Quixall and A. Bradshaw, "Phase Transformations", Institution of Metallurgists, April 1979, Series 3, Number 11, Volume 2, p.11-67.

9.9 A.J. Philips, Trans AIME, 1930, Vol 89, p 194.

9.10 T.B. Massalski, Acta Met, 1958, Vol 6, p 243.

9.11 E.A. Wilson, PhD Thesis, 1965, University of Liverpool.

9.12 W.S. Owen and E.A. Wilson, "Physical Properties of Martensite and

Bainite", ISI Special Report No 93, 1965, p 53.

9.13 R.I. Eaton, Met Obrabotka, 1956, Vol 9, p 10.

9.14 S.K. Bhattacharyya, J.H. Perepezko and T.B. Massalski, Scripta Met, 1973, Vol 7, p 485.

9.15 W.D. Swanson and J. Gordon Parr, JISI 1964, Vol 202, p 104

9.16 O.P. Marozov, D.A. Mirzayev and M.M. Shteynberg, Physics of Metals and Metallography, 1971, Vol 32(6), p 170.

9.17 B. Lee et al. Metal Science 1977, Vol 11(7), p 261.

9118 B. Cantor et al, Modern Metallography Conference, September 1980, Sheffield University.

9.19 W.S. Owen, E.A. Wilson and T. Bell, "High Strength Materials", Second International Materials Symp., University of California, 1964, Editors V.F. Zackay and H.I. Aaronson, John Wiley and Sons, New York, p 167.

9.20 "The Kinetics of Phase Transformations in Metals", J. Burke, Pergamon 1965, p 157.

9.21 S.K. Bhattacharyya, J.H. Perepezko and T.B. Massalski, Acta Met. 1974, Vol 22, p 879.

9.22 E.A. Wilson, Scripta Met, 1978, Vol 12, p 961.

9.23 K.C. Russell, Acta Met., 1969, Vol 17, p 1123.

9.24 D.P. Morozov, D.A. Mirzayev and M.M. Shteynberg, Physics of Metals and Metallography 1972, 34(4) p 114.

Footnote

Following a suggestion by Christian, more recent calculations of the kinetics of the massive transformation by the author suggest it is controlled by growth.

The incubation period for any transforamtion is given by:-

(a) The time required for nucleation, given by equation 7.17

$$\tau = \frac{\delta^2}{2\beta^*} = t_{(a)}$$

(b) The time required for the new phase to become physically observable, $t_{(b)}$.

Using equation 9.4, and assuming (b) occurs for 5% growth then it is found

$$t_{(a)} << t_{(b)}$$

A plateau is also found for the variation of $t_{(b)}$ with the degree of under-cooling at 800°C in agreement with the Russian work on zone refined iron (9.24). Also calculation of the critical cooling rate required to suppress the equiaxed ferrite transformation is in good agreement with that observed in practice (9.24).

10. Pro-eutectoid Ferrite

10.1 Morphology of Pro-eutectoid Ferrite (after Chadwick 10.1)

Two morphologies of pro-eutectoid ferrite are observed in Fe-C alloys and plain carbon steels:-

(a) Grain boundary ferrite, forming at prior austenite grain boundaries at high reaction temperatures.

(b) Widmanstatten ferrite plates forming from austenite grain boundaries and within grains at large degrees of undercooling.

Ferrite nucleates on prior austenite grain boundaries, presumably as a semi-spherical cap with one interface coherent with one austenite grain and incoherent with the other, figure 10.1.

At high reaction temperatures the incoherent interface will be highly mobile and growth will occur by movement of this interface to give an approximately equi-axed structure.

At large degrees of undercooling and lower temperatures the incoherent interface will be less mobile and growth will occur by movement of the semi-coherent interface. This occurs by lateral movement of ledges and produces surface relief (ie involves shear) and is therefore favoured by large degrees of undercooling when the chemical driving force (ie the force required to produce shear) is large. Step wise growth of plates is in fact observed in the thermionic electron emission microscope(10.2) and the overall rate is controlled by volume diffusion of carbon to the ledges. The kinetics of this process are complex (10.3) and will not be discussed further.

Instead the kinetics of grain boundary ferrite formation will be discussed, since Zener (10.4) has provided a simple treatment of this.

10.2 Rate of Thickening of Grain Boundary Allotriomorph

Consider a plane interface advancing a rate dx/dt into austenite. We can approximate the distribution of carbon to that in figures 10.2 and 10.3.

The flux of carbon across austenite/ferrite interface

$$= -D \frac{dc}{dx} = D \frac{\Delta C}{\Delta x} = D \frac{(C_\gamma - C_o)}{\Delta x}$$

Suppose the interface advances dx in tine dt, then the flux of carbon during this time interval

$$= D \frac{(C_\gamma - C_o)}{\Delta x} dt$$

This must equal the amount of carbon crossing the interface as it

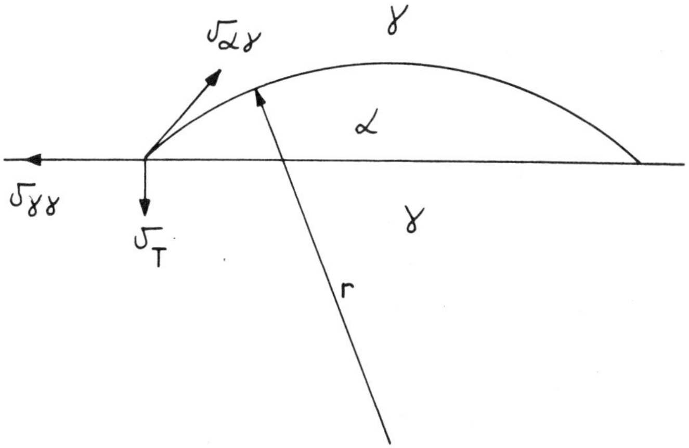

FIG. 10.1 Model of nucleus for grain boundary ferrite

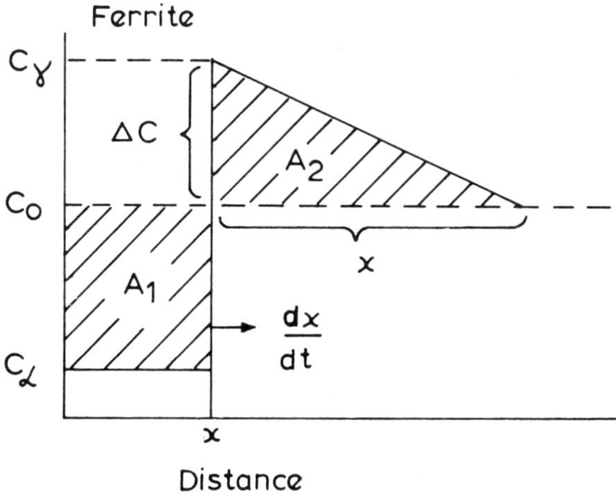

FIG. 10.2 Distribution of carbon ahead of a ferrite plate.

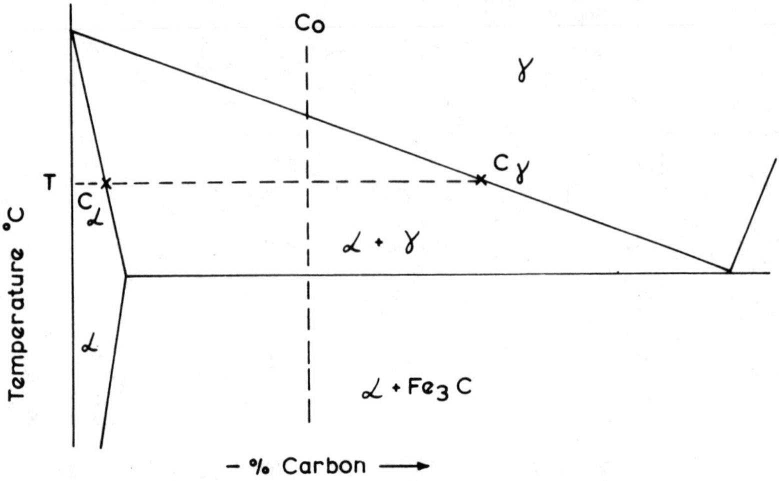

FIG. 10.3 Carbon concentration during formation of pro-eutectoid ferrite at temperature T

advances dx = $(C_\gamma - C_\alpha)$dx.

Hence for mass balance

$$(C_\gamma - C_\alpha)\frac{dx}{dt} = D\frac{\Delta C}{\Delta x} = D\frac{(C_\gamma - C_o)}{\Delta x}$$

Now the two areas A_1 and A_2 must be equal in order that the amount of carbon removed from the ferrite equals the amount of carbon diffusing into the austenite

$$\therefore \quad (C_o - C_\alpha)x = \tfrac{1}{2}\Delta C.\Delta x$$

$$= \tfrac{1}{2}(C_\gamma - C_o)\Delta x$$

Hence substituting for Δx

$$(C_\gamma - C_\alpha)\frac{dx}{dt} = D\frac{(C_\gamma - C_o)}{\Delta x} = \frac{D}{2x}\frac{(C_\gamma - C_o)^2}{(C_o - C_\alpha)}$$

Integrating $\int 2x dx = \int D\alpha dt$

$$x^2 = D\alpha t + \text{Constant}$$

$$\text{or } x = (D\alpha)^{\frac{1}{2}}t^{\frac{1}{2}} + \text{Constant}$$

When x = 0, t = 0 $\therefore$ Constant = 0

$$x = (D\alpha)^{\frac{1}{2}}t^{\frac{1}{2}}$$

$$\text{where } \alpha = \frac{(C_\gamma - C_o)^2}{(C_\gamma - C_\alpha)(C_o - C_\alpha)}$$

x = half thickness of grain boundary allotriomorph.

Similar kinetics will apply to the lengthening of a grain boundary allotriomorph.

Problem 10.1 The following results were obtained (10.5) for the lengthening kinetics of grain boundary ferrite in a 0.23% C. high purity Fe-C alloy reacted at 775°C.

Time at 775°C, seconds	4	8	12	16	24	32
Half length of allotriomorph, µm	1.75	3.25	4.75	6.20	8.00	8.25

Show that the data conforms to Zener's parabolic growth rate equation and determine the value of ($D\alpha$) and hence D corresponding to this data.

Half length of the allotriomorph L/2 should vary as $t^{\frac{1}{2}}$ and the plot in figure 10.4, shows that this is the case. Note the incubation period, t_c.

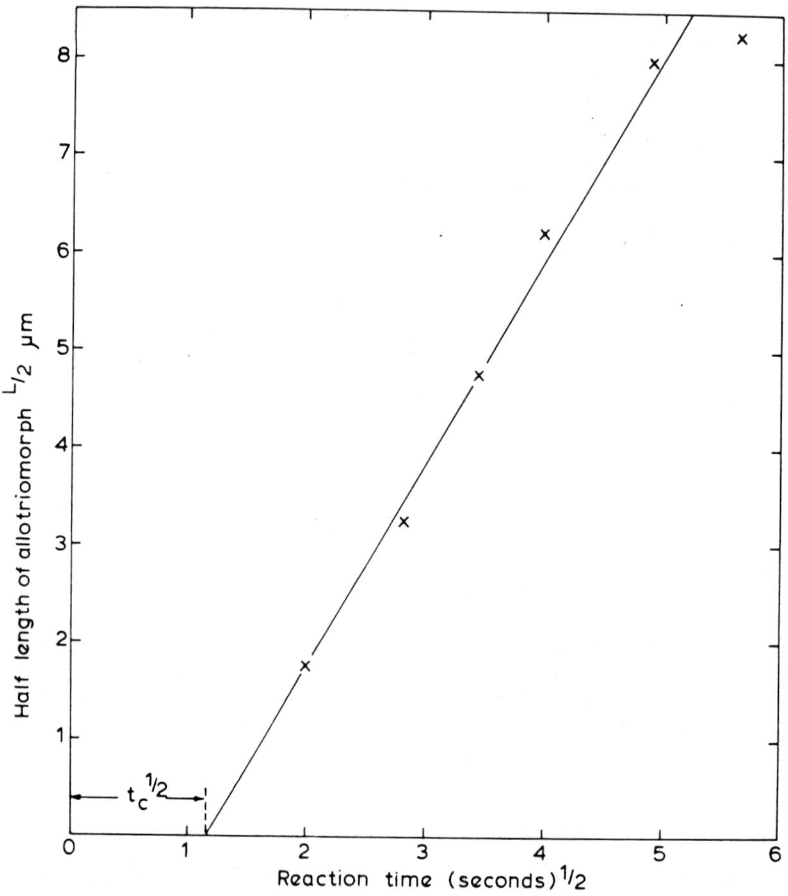

FIG. 10.4 Variation of lengthening kinetics of grain boundary ferrite with time in an Fe - 0.23 % C alloy.

Slope of this line $(D\alpha)^{\frac{1}{2}} = \frac{7.6 - 1.30}{4.8 - 1.8}$

$$= \underline{2.1 \ \mu m \ secs^{-\frac{1}{2}}}$$

Now C_o = 0.23%C and assuming the phase boundaries in $Fe-Fe_3C$ phase diagram are straight C_α and C_γ are given by

$$C_\alpha = 0.02 - \frac{(775 - 723)}{(910 - 723)} \times 0.02 = 0.0144 \text{ mass \%}$$

$$C_\gamma = \frac{(910 - 775)}{(910 - 723)} \times 0.8 = 0.578 \text{ mass \%}$$

$$\text{Hence} \quad = \frac{(C_\gamma - C_\alpha)^2}{(C_\gamma - C_\alpha)(C_o - C_\alpha)} = 0.997$$

Hence D at 775°C = $\underline{4.42 \ \mu m^2/sec}$

This may be compared with the value of 1.5 μm^2/sec given by the data of Wells et al (10.8).

The difference is thought to be due to the presence of some ledges on the allotriomorph which restrict its rate of movement (10.5).

10.3 Chemical Driving Force for formation of pro-eutectoid ferrite (After H.I. Aaronson 10.9)

Consider one mole containing n_o atom fraction of carbon reacting at some temperature T to form α and γ of compositions n_α and n_γ figure 10.5.

$$\gamma \longrightarrow \alpha + \gamma$$
$$\text{1 mol} \quad \text{u mols} \quad \text{v mols}$$

$$u = \frac{n_\gamma - n_o}{n_\gamma - n_\alpha} \simeq \frac{C_\gamma - C_o}{C_\gamma - C_\alpha}$$

$$v = \frac{n_o - n_\alpha}{n_\gamma - n_\alpha} \simeq \frac{C_o - C_\alpha}{C_\gamma - C_\alpha}$$

The approximations are extremely good.

Note C_α < 0.02 mass % carbon
< 0.0043 atomic %

C_γ & C_o < 0.8 mass % carbon
< 0.172 atomic %

Since n_α < 0.0043 atomic % we can neglect the contribution of carbon to the free energy of ferrite and then

$$G^\alpha \simeq G^\alpha_{Fe}$$

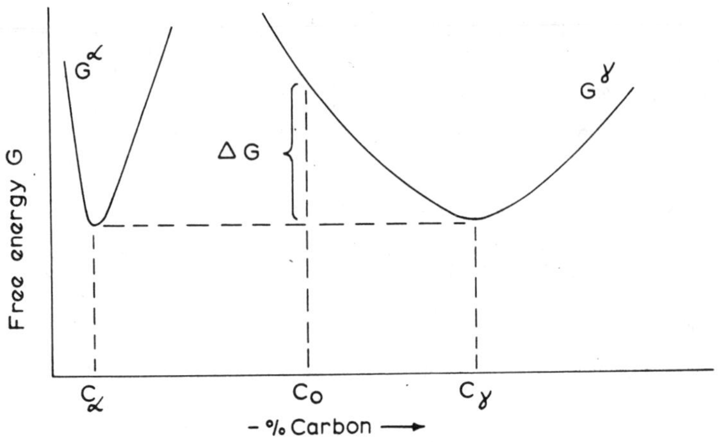

FIG. 10.5 Chemical driving force for formation pro-eutectoid ferrite.

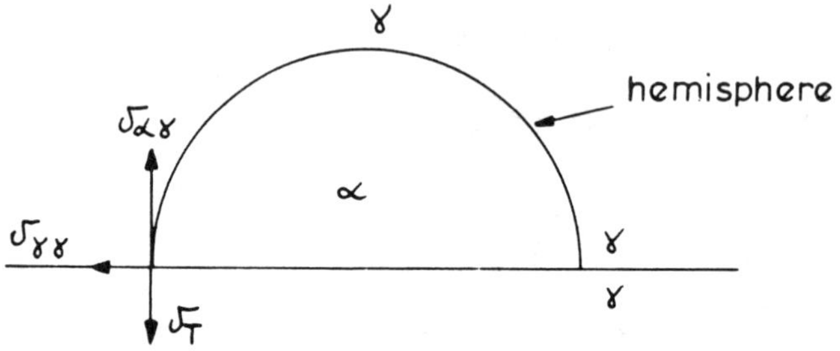

FIG. 10.6 Simplfied nucleus for pro-eutectoid ferrite.

and $G^{\gamma} = N_{Fe}G^{o}_{Fe} + N_{C}G^{o}_{C} + N_{Fe}RT\ln a_{Fe} + N_{C}RT\ln a_{C}$

$\simeq N_{Fe}G^{o}_{Fe} + N_{Fe}RT\ln a_{Fe}$ since $N_{C} < 0.172$ atomic % is small

$\simeq (1-n_{\gamma})G^{\gamma}_{Fe} + (1-n_{\gamma})RT\ln(1-n_{\gamma})$ if ideal,

$\simeq G^{\gamma}_{Fe} + RT\ln(1-n_{\gamma})$, since $n_{\gamma} < 0.172$ atomic %

Hence the total free energy G_{final} of the product phases is given by

$$\begin{aligned} G_{final} &= uG^{\alpha} + vG^{\gamma} \\ &= uG^{\alpha}_{Fe} + vG^{\gamma}_{Fe} + vRT\ln(1-n_{\gamma}) \\ &= uG^{\alpha}_{Fe} + vG^{\gamma}_{Fe} + RT\ln(1-n_{\gamma})^{v} \end{aligned}$$

Hence the chemical driving force for the formation of ferrite is given by

$$\begin{aligned} \Delta G_{v} &= G_{final} - G_{initial} \\ &= uG^{\alpha}_{Fe} + vG^{\gamma}_{Fe} + RT\ln(1-n_{\gamma})^{v} - G^{\gamma}_{Fe} - RT\ln(1-n_{o}) \\ &= uG^{\alpha}_{Fe} + vG^{\gamma}_{Fe} + RT\ln(1-n_{\gamma})^{v} - (u+v)G^{\gamma}_{Fe} - RT\ln(1-n_{o}) \\ &= u(G^{\alpha}_{Fe} - G^{\gamma}_{Fe}) + RT\ln\frac{(1-n_{\gamma})^{v}}{(1-n_{o})} \\ &= u\Delta G^{\gamma\to\alpha}_{Fe} + RT\ln\frac{(1-n_{\gamma})^{v}}{(1-n_{o})} \end{aligned}$$

where ΔG_{Fe} = Free energy change for $\gamma\to\alpha$ for one mol of pure iron at temperature T

ΔG is the chemical driving force/mol of austenite

If Ω = molar volume of austenite, then

$$\Delta G_{v} = \frac{u}{\Omega}\Delta G^{\gamma\to\alpha}_{Fe} + \frac{RT}{\Omega}\ln\frac{(1-n_{\gamma})^{v}}{(1-n_{o})}$$

= chemical driving force/unit volume of austenite.

Problem 10.2 Calculate the chemical driving force for the formation of pro-eutectoid ferrite at 860°C and 730°C in mild steel containing 0.08%C.

Data

Molar volume of austenite at 850°C = 7.26 ml

Temperature K =	1183	1180	1160	1140	1120	1100
$\Delta G^{\gamma\to\alpha}_{Fe}$ J/mol =	0	-4	-17	-33	-54	-84

Temperature K = 1080   1060   1040   1020   1000

$\Delta G_{Fe}^{\gamma \to \alpha}$ J/mol = -121   -163   -209   -264   -318

Relative atomic mass of Fe = 55.85

Relative atomic mass of C = 12.01

It will be assumed that the steel is essentially an Fe-0.08%C alloy and the variation of molar volume with temperature will be neglected.

Assuming that the phase boundaries in Fe-Fe_3C phase diagram are linear, then at 860°C

$$C_\gamma = \frac{(910 - 860)}{(910 - 727)} \times 0.8 = 0.214 \text{ mass\%}$$

$$= 0.214 \times \frac{55.85}{12.05} \times \frac{1}{100} = 0.009952 \text{ mol fraction } (n_\gamma)$$

$$C_\alpha = \frac{(910 - 860)}{(910 - 723)} \times 0.02 = 0.00535 \text{ mass\%}$$

$$C_o = 0.08 \text{ mass\%} = 0.08 \times \frac{55.85}{12.01} \times \frac{1}{100} = 0.0037202 \text{ mol fraction } (n_o)$$

$$= \frac{C_\gamma - C_o}{C_\gamma - C_\alpha} = \frac{0.214 - 0.08}{0.214 - 0.00535} = 0.6422$$

$$v = \frac{C_o - C_\alpha}{C_\gamma - C_\alpha} = \frac{0.08 - 0.00535}{0.214 - 0.00535} = 0.3578 = (1-u)$$

At 860°C = 1133K $\Delta G_{Fe}^{\gamma \to \alpha} = 54 - \frac{(54 - 33)}{(1140 - 1120)} \times (1133 - 1120)$

= 40.35 J/mol by interpolation of given data

Hence chemical driving force at 860°C, ΔG_v

$$= \frac{u}{\Omega} \Delta G_{Fe}^{\gamma \to \alpha} + \frac{RT}{\Omega} \ln \frac{(1 - n_\gamma)^v}{(1 - n_o)}$$

$$= - \frac{0.6422}{7.26} \times 40.35 + \frac{8.314 \times 1133}{7.26} \ln \frac{(1 - 0.009952)^{0.3578}}{(1 - 0.00372)}$$

$$= \underline{\underline{-3.38 \text{ J/ml}}}$$

At 730°C $C_\gamma = \frac{(910 - 730)}{(910 - 723)} \times 0.8 = 0.77$ mass%

$$= \frac{0.77 \times 55.85}{100 \times 12.01} = 0.03581 \text{ mol fraction } (n_\gamma)$$

$$C_\alpha = \frac{(910 - 730)}{(910 - 723)} \times 0.02 = 0.01925 \text{ mass\%}$$

$$C_o = 0.08 \text{ mass\%} = 0.003720 \text{ mol fraction } (n_o)$$

$$u = \frac{C_\gamma - C_o}{C_\gamma - C_\alpha} = \frac{0.77 - 0.08}{0.77 - 0.01925} = 0.91908$$

$$v = \frac{C_o - C_\alpha}{C_\gamma - C_\alpha} = \frac{0.08 - 0.01925}{0.77 - 0.01925} = 0.08092 = (1 - u)$$

At $730^\circ C = 1003K$, $\Delta G_{Fe}^{\gamma\to\alpha} = -318 + \frac{3}{20} \times (318 - 264)$

$= -309.9$ J/mol by interpolation of given data

Hence chemical driving force at $730^\circ C$, ΔG_v

$$= \frac{U}{\Omega} \Delta G_{Fe}^{\gamma\to\alpha} + \frac{RT}{\Omega} \ln \frac{(1 - n_\gamma)^v}{(1 - n_o)}$$

$$= \frac{0.91908}{7.26} \times -309.9 + \frac{8.314 \times 1003}{7.26} \ln \frac{(1 - 0.03581)^{0.08092}}{(1 - 0.00372)}$$

$$= \underline{\underline{-38.34 \text{ J/ml}}}$$

10.4 Nucleation of Pro-eutectoid Ferrite

For the nucleus shown in figure 10.6

$$\Delta G = 2/3\pi r^3 (\Delta G_v + W) + \pi r^2 \sigma_{\gamma\alpha} + 2\pi r^2 \sigma_{\gamma\alpha} - \pi r^2 \sigma_{\gamma\gamma}$$

$$= 2/3\pi r^3 (\Delta G_v + W) + 2\pi r^2 \sigma_{\gamma\alpha} + \pi r^2 (\sigma_{\gamma\alpha} - \sigma_{\gamma\gamma})$$

$$= 2/3\pi r^3 (\Delta G_v + W) + \pi r^2 (3\sigma_{\gamma\alpha} - \sigma_{\gamma\gamma})$$

Clearly, this is not the equilibrium nucleus shape, since if it was we would have,resolving the surface energy terms,

$\sigma_T = \sigma_{\gamma\alpha}$ and $\sigma_{\gamma\alpha} = \sigma_{\gamma\gamma}$

which in general is not the case.

The equilibrium shape is in fact a semi-sphere as shown in figure 10.1. However for ease of calculation we will assume that the nucleus is a hemisphere. This will not introduce appreciable error.

At ΔG^*

$$\frac{d(\Delta G)}{dr} = 2\pi r^2 (\Delta G_v + W) + 2\pi r (3\sigma_{\gamma\alpha} - \sigma_{\gamma\gamma}) = 0$$

$$\therefore \quad 2\pi r^2 (\Delta G_v + W) = -2\pi r (3\sigma_{\gamma\alpha} - \sigma_{\gamma\gamma})$$

$$\therefore \quad r^* = -\frac{(3\sigma_{\gamma\alpha} - \sigma_{\gamma\gamma})}{\Delta G_v + W}$$

Hence substituting in the original equation for ΔG

$$\Delta G^* = -2/3\pi \frac{(3\sigma_{\gamma\alpha} - \sigma_{\gamma\gamma})^3}{(\Delta G_v + W)^3}(\Delta G_v + W) + \pi \frac{(3\sigma_{\gamma\alpha} - \sigma_{\gamma\gamma})^2}{(\Delta G_v + W)^2}(3\sigma_{\gamma\alpha} - \sigma_{\gamma\gamma})$$

$$= \frac{\pi}{3} \frac{(3\sigma_{\gamma\alpha} - \sigma_{\gamma\gamma})^3}{(\Delta G_v + W)^2}$$

$\therefore$ Rate of nucleation $I = K \exp - \{1/3\pi \frac{(3\sigma_{\gamma\alpha} - \sigma_{\gamma\gamma})^3}{(\Delta G_v + W)^2} + Q\}/RT$

where R = activation energy for diffusion of carbon in austenite.

Problem 10.3 Calculate the critical nucleus size r^* for pro-eutectoid ferrite forming at $860^\circ C$ and $730^\circ C$

Data $\sigma_{\gamma\gamma} = 0.85\ J/m^2$ (10.10) $\sigma_{\gamma\gamma} = 0.56\ J/m^2$

From previous problem 10.2

At $860^\circ C$ $\Delta G_v = -3.38\ J/cc = -3.38 \times 10^6 J/m^3$

At $730^\circ C$ $\Delta G_v = -38.34\ J/cc = -38.34 \times 10^6 J/m^3$

Neglecting strain energy W

$$r^* = -\frac{(3\sigma_{\gamma\alpha} - \sigma_{\gamma\gamma})}{\Delta G_v}$$

At $860^\circ C$ $r^* = \frac{(3 \times 0.56 - 0.85)}{3.38 \times 10^6} - 2.46 \times 10^{-7} m$

$= \underline{\underline{2,460 \text{Å}}}$

At $730^\circ C$ $r^* = \frac{(3 \times 0.56 - 0.85)}{38.34 \times 10^6} - 2.16 \times 10^{-8} m$

$= \underline{\underline{216 \text{Å}}}$

The nucleus sizes are rather large compared with what can actually be expected. It is possible that nuclei at grain boundary junctions (10.10) or pill-box nuclei (10.7) may give more reasonable values.

References

10.1 G.A. Chadwick, "Metallography of Phase Transformations", London, Butterworths, 1972, pp 223-238.

10.2 K.R. Kinsman, E. Eichen and H.E. Aaronson, Met. Trans., 1975, Vol 6A, pp 303-317.

10.3 G.J. Jones and R. Trivedi, J.Appl. Phys., 1971, Vol 42, p 4299.

10.4 C. Zener, J.Appl.Phys, 1949, Vol 20, p 950.

10.5 J.R. Bradley, J.M. Rigsbee and H.I. Aaronson, Met.Trans.,1977, Vol 8A, pp 323-333.

10.6 P.J. Clemm and J.C. Fisher, Acta Met., 1955, Vol 3, pp 70-73.

10.7 W. Lange and H.I. Aaronson to be published, Met. Trans, Series A.

10.8 C. Wells, W. Bate and R.F. Mehl, Trans. AIME 1950, Vol 188, p 553.

10.9 H.I. Aaronson, "Decomposition of Austenite by Diffusional Processes", edited by V.F. Zackay and H.I. Aaronson, Interscience Publishers, 1962, pp 387-548.

10.10 P.J. Clemm and J.C. Fisher, Acta Met., 1955, Vol 3, pp. 70-73.

11. The Pearlite Transformation

The pearlite transformation occurs in plain carbon steels over a temperature range $\sim 200^{o}C$ below the eutectoid temperature. It consists of a lamellar structure of cementite and pearlite, the spacing of which, S, depends on the transformation temperature. Either ferrite or cementite can form first in the formation of pearlite, but the nucleus of pearlite really requires two adjacent nuclei of both ferrite and cementite in order to form the lamellar structure. The pearlite then grows by edgewise growth of the lamellae.

In alloy steels containing very strong carbide formers the pearlite reaction is replaced by the interphase reaction (11.1).

We will first consider the nucleation of pearlite, then its growth and finally the variation of lamellar spacing S with temperature.

11.1 Nucleation of pearlite

The following derivation is due to Fisher (11.2). Pearlite nucleates on the prior austenite grain boundaries. We will assume that cementite nucleates first at these boundaries. This depletes the surrounding austenite of carbon and ferrite nucleates on the cementite forming a nucleus for pearlite. The reverse argument applies if ferrite forms first and indeed gives a similar result to the following derivation.

Let $\dot{n}_C$ = rate of nucleation of cementite per unit area of austenite grain boundary

$\dot{n}_F$ = rate of nucleation of ferrite per unit area of cementite-austenite interface

Consider first the nucleation of ferrite on the surface of a cementite particle nucleated at time t. We can expect that the particle grows by the usual parabolic relationship and its radius r after time t is:-

$$r = \alpha t^{\frac{1}{2}}$$

and its area $A = 4\pi r^2 = 4\pi\alpha^2 t$

Hence the rate of nucleation on the cementite particle

$\dot{n} = \dot{n}_F . A = 4\pi\alpha^2 t\dot{n}_F$.

Let p be the probability that the first ferrite nucleus has formed on the cementite particle, ie the probability that pearlite has nucleated, then it can be shown by similar arguments to that for extended volumes considered in section 4.1

$$\frac{dp}{dn} = (1 - p)$$

ie $$dp = (1 - p)\frac{dn}{dt}.dt = (1 - p)\dot{n}dt$$

$$= (1 - p)4\pi\alpha^2 t\dot{n}_F dt$$

On integration

$$p = 1 - \exp(-mt^2)$$

where $$m = 2\pi\alpha^2\dot{n}_F$$

The rate of nucleation of pearlite therefore is,

$$\dot{n}_p = \frac{dp}{dt} = 2mt\ \exp(-mt^2)$$

per cementite particle.

Suppose a cementite particle forms at time τ, then time of growth $= (t - \tau)$ and $\dot{n}_c d\tau$ cementite nuclei form in the time interval $(\tau, \tau + d\tau)$.

Then rate of nucleation of pearlite at these cementite particles at time t

$$d\dot{n}_p = 2m(t - \tau)\exp\left[-m(t - \tau)^2\right]\dot{n}_c d\tau$$

Then the pearlite nucleation rate corresponding to isothermal transformation time t,

$$\dot{n}_p = \int_{\tau=0}^{\tau=t} d\dot{n}_p = \dot{n}_c\left[1 - \exp(-mt^2)\right] \quad \ldots\ldots\ldots\ldots \quad 11.1$$

Now $$e^x = 1 + x + \frac{x^2}{2!} + \frac{x^3}{3!} + \frac{x^4}{4!} + \ldots \frac{x^n}{n!}$$

$$\therefore \quad e^{-mt^2} = 1 - mt^2 + \frac{m^2t^4}{2!} - \frac{m^3t^6}{3!}$$

$$= 1 - mt^2 \text{ at small values of t.}$$

$\therefore$ Early in the transformation, rate of nucleation of pearlite,

$$\dot{n}_p = \dot{n}_c mt^2 \quad \ldots\ldots\ldots\ldots \quad 11.2$$

Problem 11.1

The following nucleation data were obtained for a eutectoid steel reacted at 680°C (11.3).

Reaction time, secs	4	5	6	6.5	8	9	10
Nucleation rate, nuclei/mm^3 sec	25 80	125 87.5	150	80 195	180 200 250	245 295 355	312.5 430 612.5

Determine whether the data obeys a parabolic relationship with time according to Fisher's formula.

Fisher's relationship is given by

$$\dot{n}_p = \dot{n}_c m t^2$$

ie $\log \dot{n}_p = \log \dot{n}_c m + 2 \log t$

Therefore a plot of $\log \dot{n}_p$ versus $\log t$ should be linear with a slope of 2.

It will be seen that this is the case within the scatter of the data, figure 11.1.

11.2 Growth of Pearlite

There are four expressions for the rate of growth of pearlite, G, assuming volume diffusion of carbon in austenite:-

$$G \alpha (T_e - T)^2 \exp - \frac{Q}{RT} \text{ due to Zener (11.4)} \quad \ldots\ldots\ldots\ldots\ldots\ldots \quad 11.3$$

$$G \alpha \frac{(C_f^\gamma - C_c^\gamma)}{(C_f^\gamma - C_f^\alpha)} (T_e - T) \exp - \frac{Q}{RT} \text{ also due to Zener (11.4)} \quad \ldots\ldots \quad 11.4$$

$$G \alpha (T_e - T) \Delta G_v^o \exp - \frac{Q}{RT} \text{ due to Frye et al (11.5)} \quad \ldots\ldots\ldots\ldots \quad 11.5$$

$$G \alpha \frac{\Delta G_v^o}{T} \exp - \frac{Q}{RT} \text{ due to Mehl and Hagel (11.6)} \quad \ldots\ldots\ldots\ldots \quad 11.6$$

C_f^γ, C_c^γ and C_f^α are defined in figure 11.2, clearly

$$\frac{(C_f^\gamma - C_c^\gamma)}{(C_f^\gamma - C_f^\alpha)} \; \alpha (T_e - T) \text{ ie } \Delta T$$

Also we can write, the chemical driving force for the transformation, ΔG_v as,

$$\Delta G_v^o = \Delta H_v^o \frac{(T_e - T)}{T_e} = \frac{\Delta H_v^o \Delta T}{T_e} \text{ in equation 11.5}$$

So expressions 11.3, 11.4 and 11.5 are equivalent:-

$$G \; \alpha \; \Delta T^2 \exp - \frac{Q}{RT} \quad \ldots\ldots\ldots\ldots\ldots\ldots\ldots\ldots\ldots\ldots\ldots\ldots \quad 11.3$$

We will only derive expressions 11.3 and 11.4 due to Zener, readers are referred to the original references for the other expressions.

At temperature T, the quasi-equilibrium concentrations will be given as in figure 11.2.

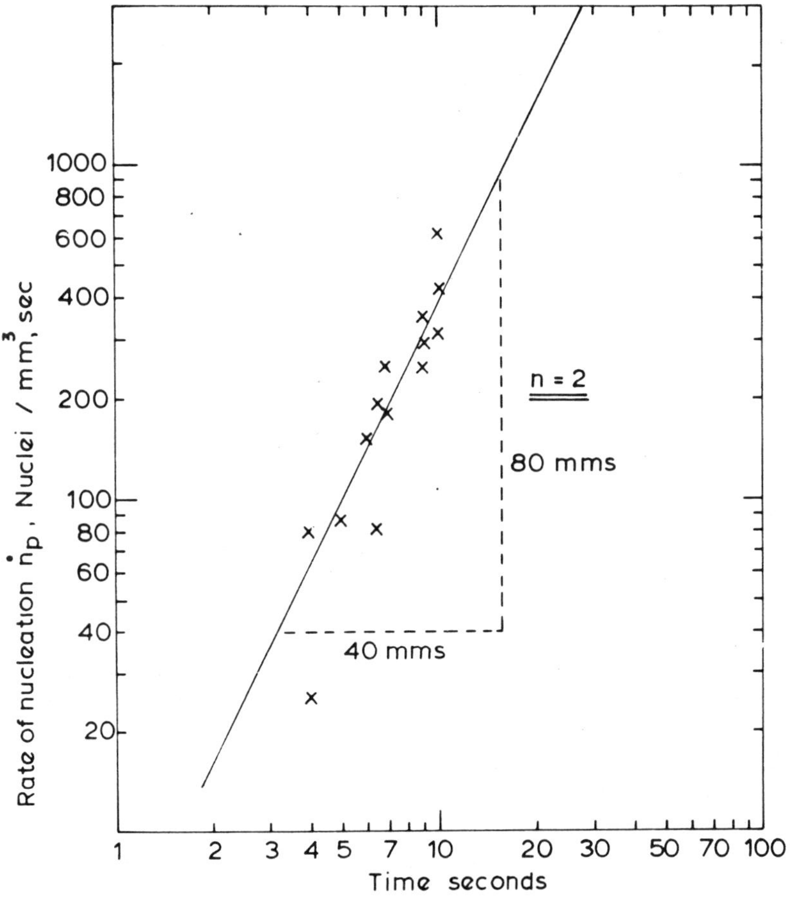

FIG.11.1 Problem 11.1 Logarithmic plot of pearlite nucleation data.

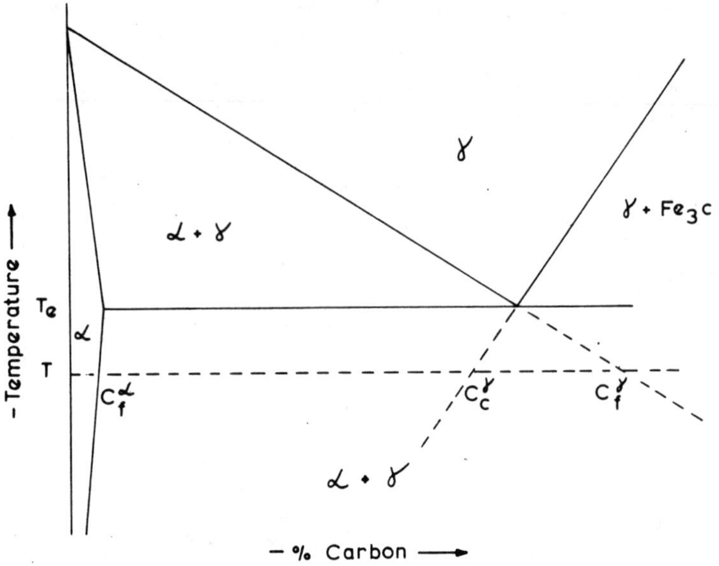

FIG. 11.2 Hultgren extrapolation (13.7) giving compositions of ferrite and austenite in contact with pearlitic ferrite and cementite (schematic)

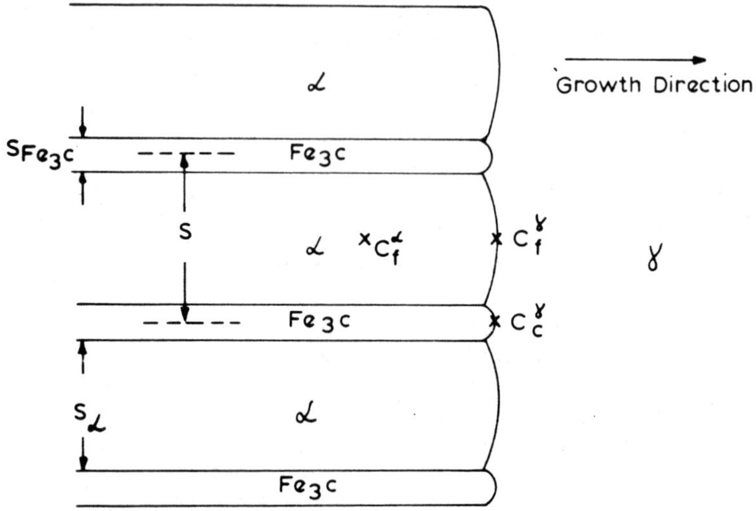

FIG. 11.3 Schematic diagram of lamellar pearlite.

From Fick's 1st Law, the flux of carbon atoms to the cementite plate J,

$$J = -D \frac{\partial c}{\partial x} = D \frac{(C_f^\gamma - C_c^\gamma)}{\alpha S}$$

Because diffusion occurs over a distance proportional to S, with $\alpha \sim \frac{1}{2}$ and the concentration gradient varies from C_f^γ at the middle of the ferrite lamella at the α/γ interface, to C_c^γ at the middle of the cementite lamella at the Fe_3C/γ interface, see figure 11.3.

Thus in time dt, the amount of carbon which has diffused/unit area =

$$D \frac{(C_f^\gamma - C_c^\gamma)}{\alpha S} dt$$

If the ferrite extends a distance dx in time dt into the austenite then the loss of carbon in the ferrite =

$$(C_f^\gamma - C_f^\alpha)\ dx$$

Hence for mass balance

$$(C_f^\gamma - C_f^\alpha)\ dx = D \frac{(C_f^\gamma - C_c^\alpha)}{\alpha S} dt$$

$$\therefore \text{ Rate of growth of lamella } = \frac{dx}{dt} = \frac{D}{\alpha S} \frac{(C_f^\gamma - C_c^\gamma)}{(C_f^\gamma - C_f^\alpha)}$$

Now Zener showed that $S \propto \frac{1}{\Delta T}$ and also we see from figure 11.2,

$$\frac{(C_f^\gamma - C_c^\gamma)}{(C_f^\gamma - C_c^\alpha)} \propto \Delta T$$

Hence

$$G \propto (\Delta T)^2 D$$

$$G \propto (\Delta T)^2 \exp - \frac{Q}{RT} \text{ per volume diffusion} \quad \ldots\ldots\ldots\ldots\ldots\ldots \quad 11.3$$

In the case of boundary diffusion at the interface D is replaced by $\frac{D_B \delta}{S}$ where D_B is the boundary diffusion coefficient and δ is the boundary width.

$$G \propto (\Delta T)^2 \frac{D_B \delta}{S}, \text{ and since } S \propto 1/\Delta T,$$

$$G \propto (\Delta T)^3 \exp - \frac{Q_B}{RT} \text{ for boundary diffusion} \quad \ldots\ldots\ldots\ldots\ldots \quad 11.7$$

In plain carbon eutectoid steels, the results obtained for G indicate that the rate of growth is controlled by volume diffusion of carbon in austenite.

Alloy elements which are carbide formers slow down the rate of growth of pearlite. This is because these substitutional elements diffuse more sluggishly than carbon and have to partition between the ferrite and the cementite in the pearlite. Direct evidence for partitioning has been obtained (11.8 and 11.9) using the analytical electron microscope.

Below a certain temperature T_p alloy elements do not partition because the temperature is too low and diffusion too sluggish for this to occur. Instead the growth rate is reduced because the alloy element affects the carbon concentration gradient.

Non-carbide forming elements do not appreciably alter the growth rate of pearlite (11.10).

11.3 Variation of Interlamellar Spacing S with degree of undercooling ΔT

Theory and experiment show that for an isothermal reaction temperature T

$$S \propto \frac{1}{(T_e - T)} \propto \frac{1}{\Delta T}$$

The theory is due to Zener (11.4) and assumes that part of the chemical driving force for the reaction ΔG_v is absorbed as interfacial energy for the α/Fe_3C interface in the pearlite and the remaining accounts for the driving force for the reaction.

Thus at small degrees of undercooling ΔT, the driving force is small and there is little energy available per unit volume for the Fe_3C/α interface and hence S is large, because this results in effect in less Fe_3C/α surface per unit volume of pearlite.

At large degrees of undercooling, ΔG_v is large and the converse applies, ie there is more energy available for the Fe_3C/α interface and consequently S is small.

An additional reason for S decreasing with increasing ΔT is that carbon has to re-distribute itself between low carbon ferrite, α, and high carbon cementite, Fe_3C over a distance $\sim S/2$. As the temperature decreases diffusion becomes more sluggish and to compensate for this, S decreases.

Energy associated with α/Fe_3C interface/unit volume of pearlite

$$\Delta G_s = \frac{2\sigma}{S}$$

Hence the driving force left for transformation

$$\Delta G = \Delta G_v^o - \frac{2\sigma}{S}$$

We can define a minimum possible spacing S_c for growth when $\Delta G = 0$, such that $S_c = \frac{2\sigma}{\Delta G_v^o}$ 11.8

Hence substituting in this value of S_c

$$\Delta G = \Delta G_v^o \left(1 - \frac{S_c}{S}\right)$$

It seems reasonable to assume that the rate of growth is determined by the redistribution of carbon between ferrite and cementite over a distance proportional to S and the driving force ΔG.

$$\therefore \text{ Rate of growth } G \propto \Delta G \frac{D}{S} = K \, \Delta G \frac{D}{S}$$

$$= K \, D \frac{\Delta G_v^o}{S}\left(1 - \frac{S_c}{S}\right)$$

Substituting for ΔG

D = Diffusion coefficient of carbon

We now have to evaluate S_c. Zener makes the assumption that S takes the value corresponding to the maximum growth rate

ie when $\frac{dG}{dS} = 0$

$$= K \, D \, \Delta G_v^o \left[-S^{-2} - S_c(-2S^{-3})\right]$$

$$= K \, D \frac{\Delta G_v^o}{S^2} \left[\frac{2S_c}{S} - 1\right]$$

$\therefore$ G is a maximum when $S = 2S_c$

Substituting this result in equation 11.8

$$S = \frac{4\sigma}{\Delta G_v^o} \quad \text{...} \quad 11.9$$

Now $\Delta G^o = \Delta H^o - T\Delta S^o$ and assuming that ΔH^o and ΔS^o do not vary with temperature

$$\Delta G_v^o = \frac{\Delta H_v^o}{T_e}(T_e - T)$$

Hence, substituting for ΔG_v^o in 11.9

$$S = \frac{4\sigma T_e}{\Delta H_v^o} \times \frac{1}{(T_e - T)} \quad \text{..............................} \quad 11.10a$$

$$\text{ie} \quad T = T_e - \frac{4\sigma T_e}{\Delta H_v^o} \cdot \frac{1}{S} \quad \text{..............................} \quad 11.10b$$

Puls and Kirkaldy (11.11) assume that S_c takes the value corresponding to the maximum rate of entropy production. This occurs when the following expression is a maximum

$$\frac{d_i S}{dt} = G\left(\frac{\Delta G}{T}\right)$$

Now $$G = K\, D \frac{\Delta G_v^o}{S}\left(1 - \frac{S_c}{S}\right).$$

and $$\Delta G = \Delta G_v^o \left(1 - \frac{S_c}{S}\right)$$

$$\therefore \quad \frac{d_i S}{dt} = K\, D \frac{(\Delta G_v^o)^2}{S}\left\{1 - \frac{S_c}{S}\right\}^2$$

$$= K\, D\, (\Delta G_v^o)^2 \left\{\frac{1}{S} - \frac{2S_c}{S^2} + \frac{S_c^2}{S^3}\right\}.$$

Differentiating this expression with respect to the interlamellar spacing S gives us

$$K\, D\, (\Delta G_v^o)^2 \left\{-\frac{1}{S^2} + \frac{4S_c}{S^3} - 3\frac{S_c^2}{S^4}\right\}$$

$$= K\, D (\Delta G_v^o)^2 \frac{1}{S^2}\left(\frac{3S_c}{S} - 1\right)\left(1 - \frac{S_c}{S}\right)$$

$$= 0 \quad \text{when} \quad \underline{\underline{S = 3S_c}}$$

Corresponding to the maximum rate of entropy production.

Problem 11.2 Q.6 Associateship in Metallurgy, Physical Metallurgy I, June 1975, Sheffield Polytechnic.

Derive the Zener relationship between the interlamellar spacing of pearlite and the transformation temperature to pearlite.

Show that the following data for the pearlite transformation in a pure Fe-0.82%C alloy conform with the Zener equation and determine the equilibrium eutectoid temperature.

Transformation Temp. ^oC	717	703	691	680	667
Interlamellar spacing, Å	5,000	2,500	1,667	1,250	1,000

The derivation is given in Section 11.3.

The problem uses equation 11.10b. A plot of T versus 1/s is linear with intercept T_e.

Transformation Temp $^{\circ}C$	717	703	691	680	667
TK	990	976	964	953	940
Interlamellar spacing S,Å	5,000	2,500	1,667	1,250	1,000
1/s Å x 10^4	2	4	6	8	10

The plot is shown in figure 11.4. It will be seen that the intercept gives T_e = 1001.5K = 728.5°C. The original data (11.3) gave a value of 787°C. Note it is not necessary to convert to absolute temperature in plotting the graph, although if say σ is to be evaluated from the slope, knowing ΔH_v, T_e must be in Kelvin.

Problem 11.3 Q.2 Associateship in Metallurgy, Physical Metallurgy B, June 1977, Sheffield City Polytechnic.

An Fe-0.8%C alloy isothermally transformed to pearlite at 690°C has an interlamellar spacing of 1,540Å. From the data given calculate the widths of the ferrite and cementite lamellae in the pearlite.
If the same alloy is transformed to pearlite at 677°C, what will be the new widths of the ferrite and cementite lamellae in the pearlite?
Explain the effect of an addition of 1%Mn to the alloy on the interlamellar spacings of the pearlite at these temperatures.
Data:

Solubility of carbon in ferrite is 0.02% at 727°C

Eutectoid temperature of Fe-0.8%C alloy; 727°C

Density of iron = 7.86 g/cm^3

Density of Fe_3C = 7.4 g/cm^3

Relative atomic mass of Fe = 55.85

Relative atomic mass of C = 12.01

The relative widths of the cementite and ferrite lamellae can be found by applying the lever rule.

It is first necessary to calculate the composition of Fe_3C in mass %.

Composition of Fe_3C = 25 atomic per cent C

$$= \frac{25 \times 12.01}{(25 \times 12.01) + (75 \times 55.85)} \times 100$$

= 6.688 mass %

Relative proportions of ferrite to cementite by mass

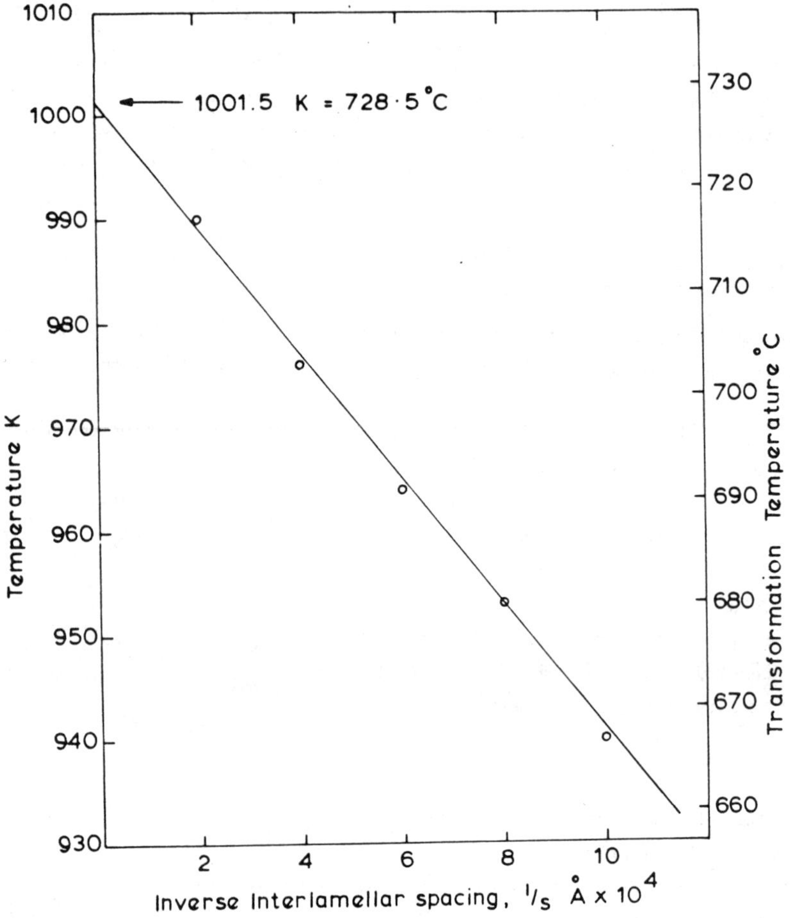

FIG. 11.4 Problem 11.2, Variation of interlamellar spacing with temperature.

$$= \frac{6.688 - 0.8}{0.8 - 0.02} = 7.55:1$$

Relative proportions of ferrite to cementite by volume

$$= 7.55 \times \frac{7.4}{7.86} = 8.108:1$$

$$\text{ie } \frac{S_\alpha}{S_{Fe_3C}} = 8.108$$

$$\text{Now } S = S_\alpha + S_{Fe_3C} = 8.108\ S_{Fe_3C} + S_{Fe_3C} = 1{,}540\text{Å}$$

$$\therefore\ S_{Fe_3C} = \frac{1{,}540}{9.108} = \underline{\underline{169\text{Å}}}$$

$$\therefore\ S_\alpha = 1{,}540 - 169 = \underline{\underline{1371\text{Å}}}$$

$$\text{Now } S \propto \frac{1}{\Delta T}, \text{ at } 690^\circ C,\ \Delta T = 727 - 690^\circ C = 37^\circ C$$

$$\therefore\ \text{At } 677^\circ C,\ \Delta T = 50^\circ C,\ \therefore\ S = 1{,}540 \times \frac{37}{50} = 1{,}140\text{Å}$$

$$\therefore\ \text{At } 677^\circ C,\ S_{Fe_3C} = \frac{1{,}140}{9.108} = \underline{\underline{125\text{Å}}}$$

$$S_\alpha = 1{,}140 - 125 = \underline{\underline{1{,}015\text{Å}}}$$

The addition of manganese will lower the eutectoid temperature, thus decreasing the degree of undercooling ΔT and increasing the lamellar spacing.

Problem 11.4 From experiments on isothermal (11.12) and iso-velocity (11.13) (pearlite transformed under an imposed moving temperature gradient), pearlite, the following values of interlamellar spacing S and growth rate V were obtained. Determine whether the data is consistent with volume or boundary diffusion of carbon.

S, cm x 10^6	180	110	110	70	51	58	34	25
G, cm/s x 10^6	4.2	12	19	43	76	120	170	430

S, cm x 10^6	21	15	7.1	6.7
G, cm/s x 10^6	490	1,200	4,100	5,500

Equations 11.3 and 11.7 give

$$G \alpha (\Delta T)^2 \exp - \frac{Q}{RT} \text{ for volume diffusion}$$

$$G \alpha (\Delta T)^3 \exp - \frac{Q_B}{RT} \text{ for boundary diffusion}$$

Now $S \alpha \frac{1}{\Delta T}$, so

$$GS^2 \alpha \exp - \frac{Q}{RT} \text{ for volume diffusion}$$

$$GS^3 \alpha \exp - \frac{Q_B}{RT} \text{ for boundary diffusion}$$

Puls and Kirkaldy (11.11) argue that over the range of temperature of measurement of the data, $\sim 100^\circ C$ that the exponential term does not vary appreciably because the carbon concentration is varying as well as the temperature. Since the activation energy also depends on the carbon concentration, calculations show that the diffusivity D, and in turn the exponential term, $\exp - \frac{Q}{RT}$, are essentially constant over the temperature range of the data.

Hence we have

GS^2 = constant, say K_v, for volume diffusion

and GS^3 = constant, say K_b, for boundary diffusion.

Taking logarithms

$$\log G + 2\log S = \log K_v$$

$$\log G + 3\log S = \log K_b$$

Therefore on a log/log plot the data will have a slope of 2 for volume diffusion and 3 for boundary diffusion, figure 11.5. It will be seen that the data is consistent with volume diffusion.

It should be noted, however, that other authors (11.14) have allowed for the exponential term by plotting $\ln GS^3$ or $\ln GS^2$ versus $1/T$. Although a linear plot is obtained with GS^3 (boundary diffusion) the activation energy obtained, Q_B = 188 kJ/mol is too large for boundary diffusion.

References

11.1 R.W.K. Honeycombe, Met. Trans., 1976, Vol. 9a, p 915.

11.2 J.C. Fisher, "Thermodynamics in Physical Metallurgy", pp 201-241, edited by C. Zener, ASM, 1949.

11.3 F.C. Hull, R.A. Colton and R.F. Mehl, Trans AIMME, 1942, 150, p 185.

11.4 C. Zener, Trans AIME, 1946, 167, p 550.

11.5 J.H. Frye, Jr., E.E. Stansbury, and D.L. McElroy, Trans AIME, 1953, 197, pp 219-224.

11.6 R.F. Mehl and W.C. Hagel: Prog. Metal Phys., 1956, 6 pp 74-134.

11.7 A. Hultgren, Trans ASM, 1947, 39, p 915.

11.8 N.A. Razik, G.W. Lorimer and N. Ridley, Acta. Met., 1974, 22 pp 1249-1258.

11.9 N.A. Razik, G.W. Lorimer and N. Ridley, Met. Trans., 1976, 7A, p 209.

11.10 A.R. Marder and B.L. Bamfitt, Met. Trans., 1976, 7A pp 902-905.

11.11 M.P. Puls and J.S. Kirkaldy, Met. Trans., 1972, 3 pp 2777-2795.

11.12 D. Brown and N. Ridley, JISI, 1969, 207, p 1232.

11.13 G.F. Bolling and R.H. Richman, Met. Trans., 1970, 1, p 2095.

11.14 N. Ridley, D. Brown and H.I. Malik, Chemical Metallurgy of Iron and Steel, ISI, 1973, p 268-271.

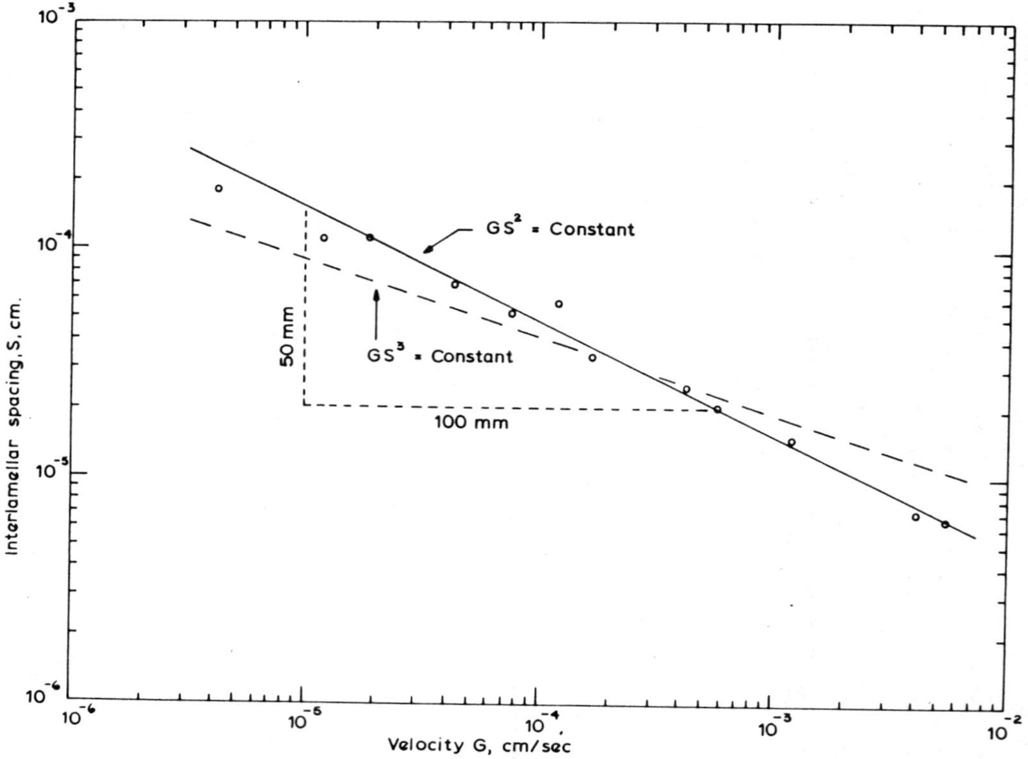

FIG. 11.5 Problem 11.4 Dependance of growth rate of pearlite on interlamellar spacing S.

12. Tempering of Martensite

12.1 Jaffe and Hollomon Tempering Parameter

These workers (12.1) investigated the variation of hardness with tempering time and temperature for a number of plain carbon steels and 2%Mo, 0.35%C secondary hardening steel.

They first assumed that the tempered hardness would be a function of the rate equation,

ie $$H = f_1(t \exp - \frac{Q}{RT}) \quad \text{12.1}$$

It was proved that the activation energy was a function of tempering time and temperature

$$Q = f_2(H) \quad \text{12.2}$$

and $$t_o = t \exp - \frac{Q}{RT} = \text{Constant} \quad \text{12.3}$$

Taking logarithms of equation 12.3

$$Q = RT(\ln t - \ln t_o) = f_2(H) \quad \text{12.4}$$

then

$$H = f_3[\exp RT\ln \frac{t}{t_o}] \quad \text{12.5}$$

ie $$= f(T \log \frac{t}{t_o})$$

$$= f[T(\log t - \log t_o)]$$

$$= f[T(20 + \log t)] \quad \text{12.6}$$

with T in Kelvin, t in hours and H in V.P.N.

The expression has been found to hold surprisingly for a number of plain carbon and alloy steels with a value of $\log t_o = -20$. It is most frequently used as a master tempering parameter, see for example figure 12.1. The scatter in this figure at low values of T(20 + logt) is due to short tempering times, $\leqslant$ 1.5 minutes at low tempering temperature, resulting in an increase in hardness (ie age hardening).

The practical value of the parameter is that it allows combination of times and temperatures to be calculated to give the same hardness, since for constant hardness

$$T_1(20 + \log t_1) = T_2(20 + \log t_2) \quad \text{12.7}$$

Problem 12.1

It takes 24 hours to achieve a hardness of 200 Hv at 600°C in a 0.31% plain carbon steel. Calculate the equivalent time to reach the same hardness

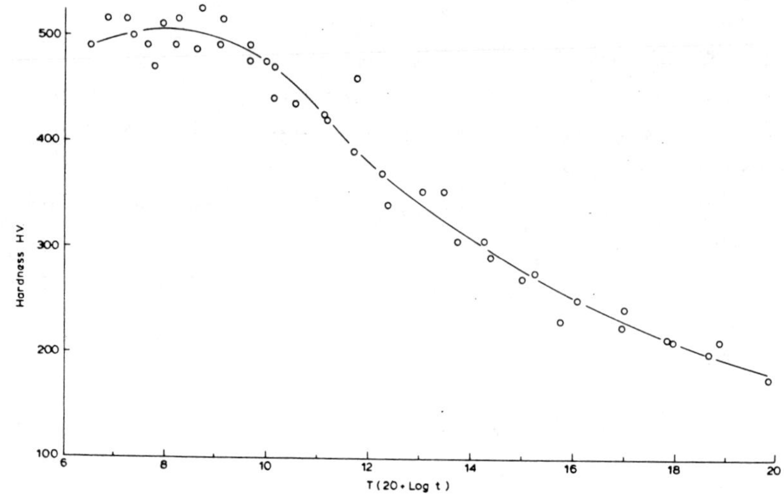

Fig. 12.1 Variation of tempered martensite hardness with Hollomon and Jaffe tempering paramter T(20 + log t) for a 0.3% plain carbon steel, tempering temperature T in Kelvin, tempering time t in hours.

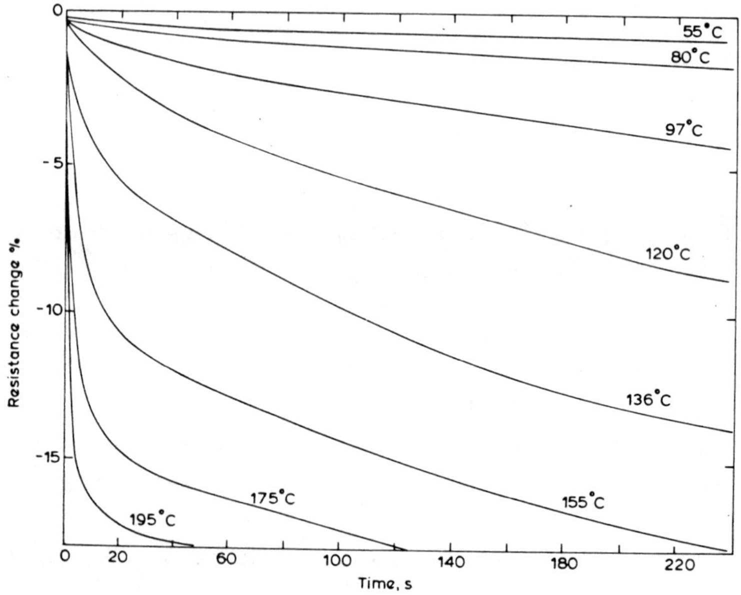

FIG. 12.2 Initial resistance changes on tempering 100% Martensite in 1% plain carbon steel.

at $700^{\circ}C$. What value is the activation energy Q to achieve this level of hardness?

$$T_1 = 600 + 273 = 873K \qquad t_1 = 24 \text{ hours}$$

$$T_2 = 700 + 277 = 973K \qquad t_2 = ?$$

$$\therefore \quad T_1(20 + \log t_1) = 18{,}665 = T_2(20 + \log t_2) \text{ from equation 12.7}$$

$$\therefore \quad \log t_2 = \frac{18{,}665}{973} - 20 = -0.817$$

$$\therefore \quad t_2 = 0.152 \text{ hours} \simeq \underline{\underline{9 \text{ minutes}}}$$

From equation 12.4

$$Q = RT(\ln t - \ln t_o) = 2.303\ RT\ (20 + \log t)$$

$$= \underline{\underline{357 \text{ kJ/mol}}}$$

The activation energy is rather high, but possibly may represent the activation energy required for recrystallisation of the martensitic matrix.

12.2 Early stages of tempering

There is some disagreement about the early stages of tempering at low temperatures < $200^{\circ}C$. Speich and Leslie (12.2) have shown by electrical resistivity techniques and internal friction that in plain carbon steels up to 0.2% carbon the carbon is trapped at dislocation sites and lath boundaries during the quench giving a b.c.c. structure. An alternative explanation of the b.c.c. structure is in terms of Zener ordering (12.3, 12.4 and 12.5). Zener postulates there is a critical temperature T_c below which ordering occurs to give a b.c.t. lattice. Thus if M_s lies below this temperature a b.c.t. lattice is obtained while if M_s is above T_c a b.c.c. lattice is obtained. Speich and Leslie (12.2) argue that above 0.2% carbon, all the sites for segregation such as lath boundaries and dislocations are filled and the remaining carbon is held in the lattice to give a b.c.t. structure.

Alternatively Kurdjunov (12.11) has shown that in high carbon martensites (∿1%C) there is an orthorhombic distortion of the martensite lattice in steels with an M_s below room temperature. On heating to room temperature the orthorhombic structure is replaced by the normal b.c.t. martensite.

F.G. Wilson (12.5 and 12.7) showed that there was a rapid (∿ seconds) decrease in resistivity on tempering 1%C plain carbon steel, see figure 12.2. He interpreted this as being due to rapid segregation of carbon to the twin boundaries within the martensite. However, at longer times King and Glover (12.12) report an increase in resistivity. They attribute this to coherency strains associated with the formation of ε carbide, although their results may be anomalous since they only measured resistivity changes after 1 minute tempering.

Japanese workers (12.8) have reported the existence of a number of intermediate carbides (four all told including ε carbide $Fe_{2.4}C$ at various tempering temperatures).

Miller et al (12.9) by a combination of FIM, atom probe microanalysis and TEM have reported the segregation of carbon to lath boundaries in an Fe-0.2%C alloy while studies on an Fe-0.8%C alloy suggests that carbon is segregated to twin boundaries within the 'twinned martensite'.

W.S. Owen and his co-workers (12.10) have examined the early stages of tempering of any Fe-1.86%C alloy by Mössbauer spectroscopy and conclude that carbon atoms cluster to form ordered regions of Fe_4C with an activation energy of 89.5 kJ/mol.

Problem 12.2

Take the data of F.G. Wilson (12.7), figure 12.2 - 12.5 and show the data in the early stages is proportional to $\sqrt{t}$.

From the slope of this plot determine the diffusion coefficient D in the martensite and hence evaluate the activation energy for diffusion of carbon in the martensite.

We will assume that (i) the thickness of twins in the martensite is h, (ii) the initial carbon content is C_o in the twin, same as the carbon content of the steel and of the parent martensite, (iii) the carbon content at the twin interface is effectively zero and (iv) and after time t the distribution of carbon within the twin is as shown in figure 12.6.

$$\text{Flux } J = -D\frac{dc}{dx} = -D\frac{C_o}{x}$$

In time dt, amount of carbon diffusing out of slab $= D\frac{C_o}{x}dt$.

In this time the interface advances dx and amount of carbon diffusing/unit area in time dt = shaded area = $\frac{1}{2}C_o dx$.

For mass balance $D\frac{C_o}{x}dt = \frac{C_o}{2}dx$

Integrating $Dt = \frac{1}{4}x^2 + \text{Constant}$

When $t = 0$, $x = 0$ $\therefore$ constant $= 0$

We will assume that the resistivity data measure the mean carbon concentration $\bar{c}$ in the twin.

This will be given by:

$$\bar{c} = 1/h\{\tfrac{1}{2}C_o x + C_o(h - 2x) + \tfrac{1}{2}C_o x\}$$

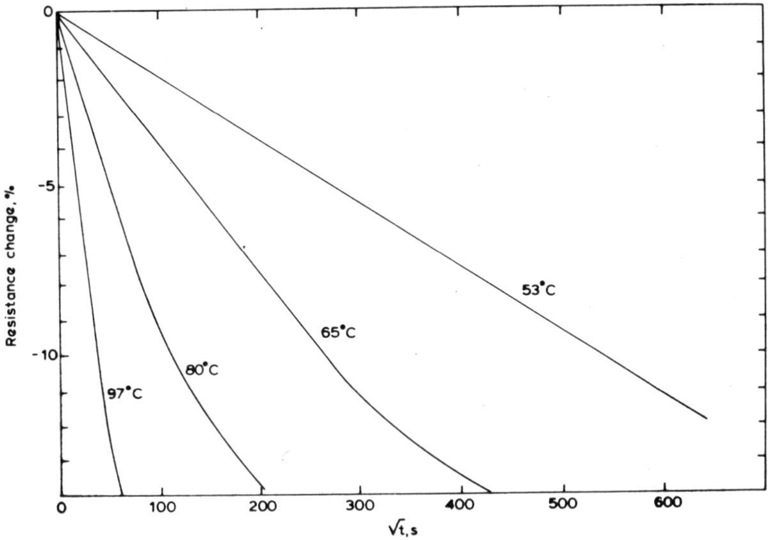

FIG. 12.3 Resistance as a function of the square root of the tempering time in seconds for 1% plain carbon steel.

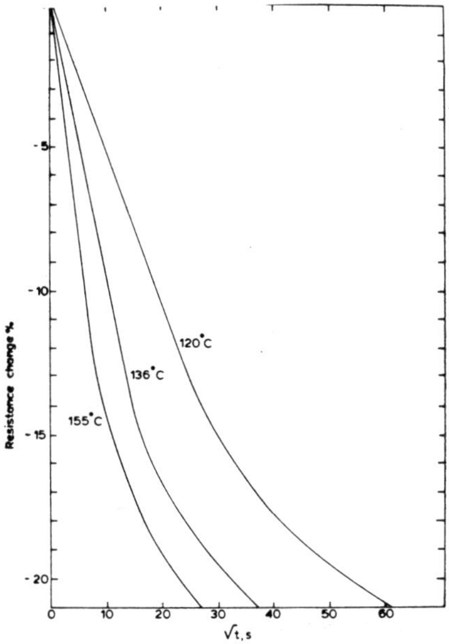

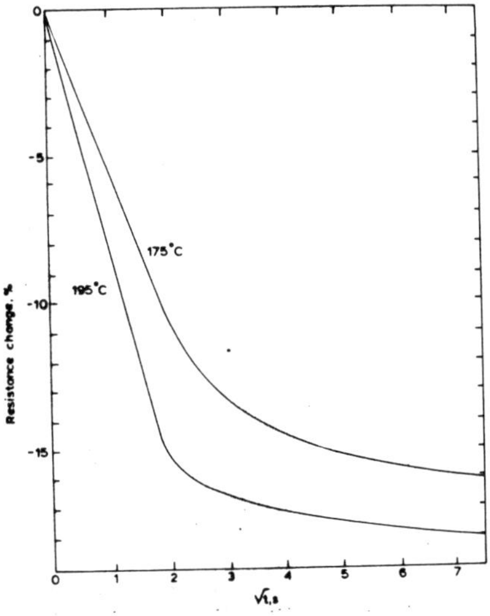

Fig. 12.4 and 12.5 Resistance as a function of the square root of the tempering time in seconds for 1% plain carbon steel.

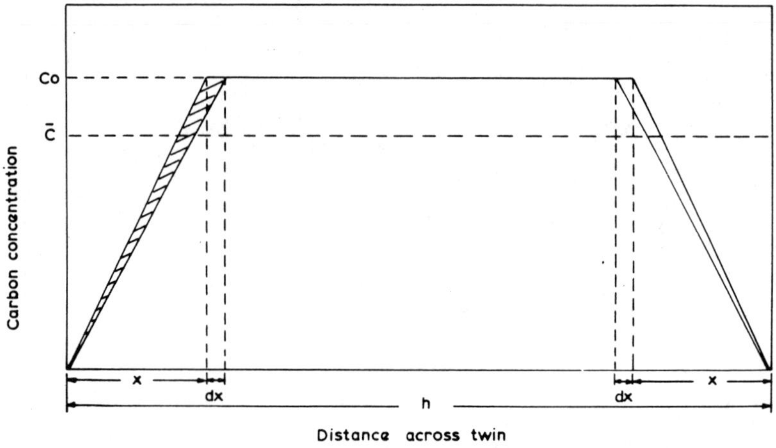

FIG. 12·6 Distribution of carbon across a twin in plate martensite.

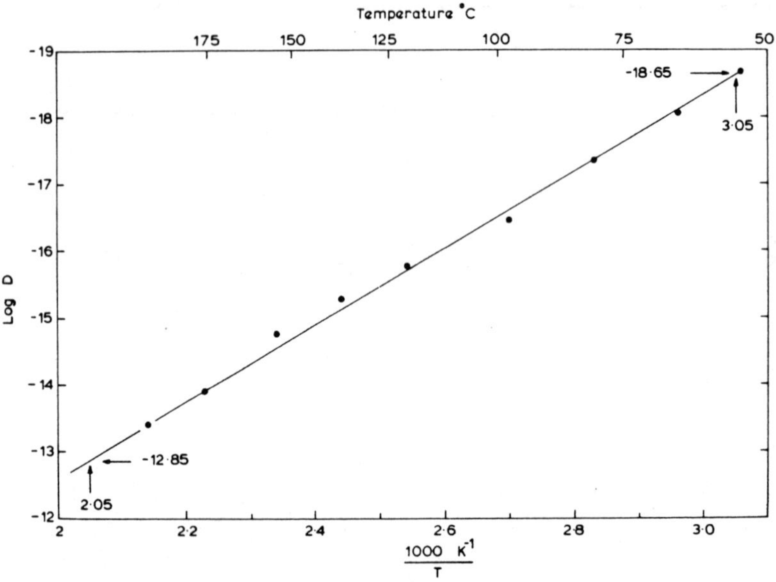

FIG. 12·7 Arrhenius plot of diffusion constant derived from early resistance changes on low tempering a 1% plain carbon steel.

$$= C_o(1 - 2\sqrt{bt}/h) \quad \text{substituting } x^2 = 4Dt$$

$$\therefore \quad \frac{\bar{C} - C_o}{C_o} = \frac{2\sqrt{bt}}{h} = \frac{R_o - R}{R} = \frac{\Delta R}{R_o}$$

Wilson and Owen's more exact analysis yields

$$\frac{\bar{C} - C_o}{C_o} = \frac{\Delta R}{R_o} = \frac{4(Dt)^{\frac{1}{2}}}{\pi^{\frac{1}{2}}h} = 2.26\,\frac{\sqrt{Dt}}{h}$$

Values of D derived from figures 12.2 to 12.5 are shown in the following table, assuming a twin thickness h = 500Å

Temperature T^oC	$\frac{1,000}{T}$ K^{-1}	$t^{\frac{1}{2}}$, $s^{\frac{1}{2}}$ at $\Delta R/Ro$ = 10%	D Cm^2/s	log D
53	3.06	557.5	2.01×10^{-19}	-18.7
65	2.96	269.3	8.62×10^{-19}	-18.06
80	2.83	113.4	4.86×10^{-18}	-17.31
97	2.70	42.06	3.53×10^{-17}	-16.45
120	2.54	18.84	1.76×10^{-16}	-15.75
136	2.44	10.46	5.71×10^{-16}	-15.24
155	2.34	6.0	1.74×10^{-15}	-14.76
175	2.23	1.97	1.61×10^{-14}	-13.79
195	2.14	1.27	3.87×10^{-14}	-13.40

From the Arrhenius plot of this diffusion data, figure 12.7

$$D = D_o \exp - \frac{Q}{RT}$$

$$\ln D = \ln D_o - \frac{Q}{RT}$$

$$\log D = \log D_o - \frac{Q}{2.303\,RT}$$

$$\therefore \quad \frac{Q}{2.303\,R} = \frac{18.65 - 12.85}{3.05 - 2.05}$$

$$Q = \underline{\underline{111 \text{ kJ/mol}}}$$

This maybe compared with the value of 106 ± 3 kJ/mol obtained by Wilson and Owen (12.7). It is too high for diffusion of carbon in ferrite, 75.4 kJ/mol. Wilson and Owen argue that the increase is due to the tetragonality of the lattice and restriction in the number of available diffusion paths.

Low carbon steels, containing less than 0.2% carbon, tend to suffer from autotempering during the quench. That is, precipitation of orthorhombic

Fe_3C occurs in the martensite below M_s during the quench (12.6). At very high cooling rates the carbon just segregates to dislocations and lath boundaries (12.10). Tempering in the range 100°C - 300°C results in little growth in the Fe_3C particles present, but the number of martensite laths containing precipitates increases. The orientation relationship is

$$(211)_{\alpha'} \; // \; (001)_{Fe_3C}$$

$$[01\bar{1}]_{\alpha'} \; // \; [100]_{Fe_3C}$$

$$[\bar{1}11]_{\alpha'} \; // \; [010]_{Fe_3C}$$

When a 0.1%C steel is tempered above 300° precipitation occurs along the lath boundaries and the cementite particles grow to rods 2000Å long x 150Å wide. After one hour at 600°C the martensite decomposition is complete and the structure consists of equiaxed ferrite grains and spheroidised grain boundary carbides. Both Wilson (12.6) and Speich (12.10) concede that ε-carbide $Fe_{2.4}C$ may form first as a transition carbide but have no evidence for this.

For steels containing more than 0.25%C three stages of tempering can be distinguished in addition to the early segregation stages described previously.

1) <u>First stage 50-200°C</u> Precipitation of ε carbide $Fe_{2.4}C$ occurs and the martensite becomes less tetragonal. An increase in hardness is observed.

2) <u>Second stage 200°C - 300°C</u> The retained austenite which is unchanged in the first stage transforms to bainite giving an increase in volume and an increase in hardness.

3) <u>Third stage</u>, above 300°C the transition carbide is converted to cementite.

These sequences are shown in figure 12.8.

For alloy steels a fourth stage of tempering can be distinguished with the precipitation of an alloy carbide at temperatures ∿500°C - 600°C to give a secondary hardening peak.

<u>Problem 12.3</u>

Dilatometry measurements gave the results in figure 12.7 from the decomposition of 100% retained austenite in an Fe-1.43%C alloy during the second stage of tempering (12.13). Determine the activation energy for this process. What does this energy correspond to?

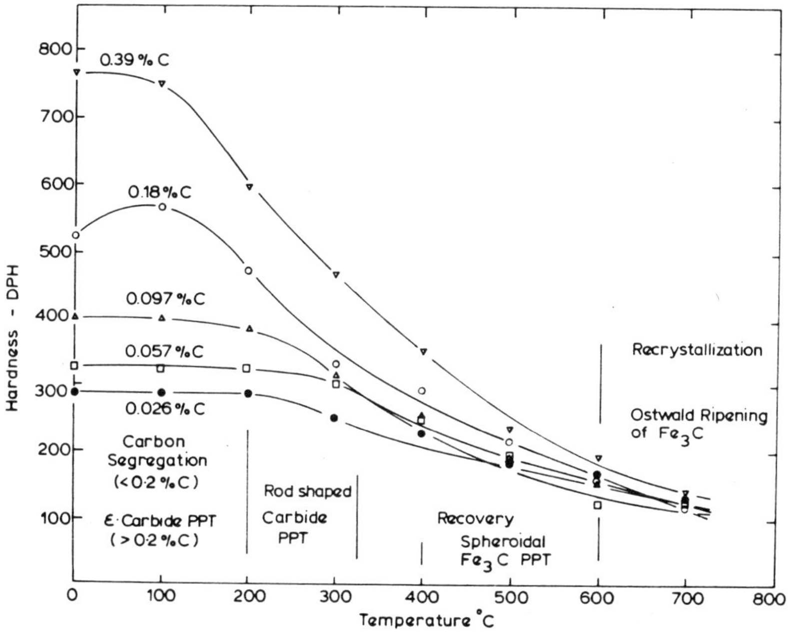

FIG. 12.8 Hardness of iron-carbon martensites tempered 1hr at 100° to 700° C.

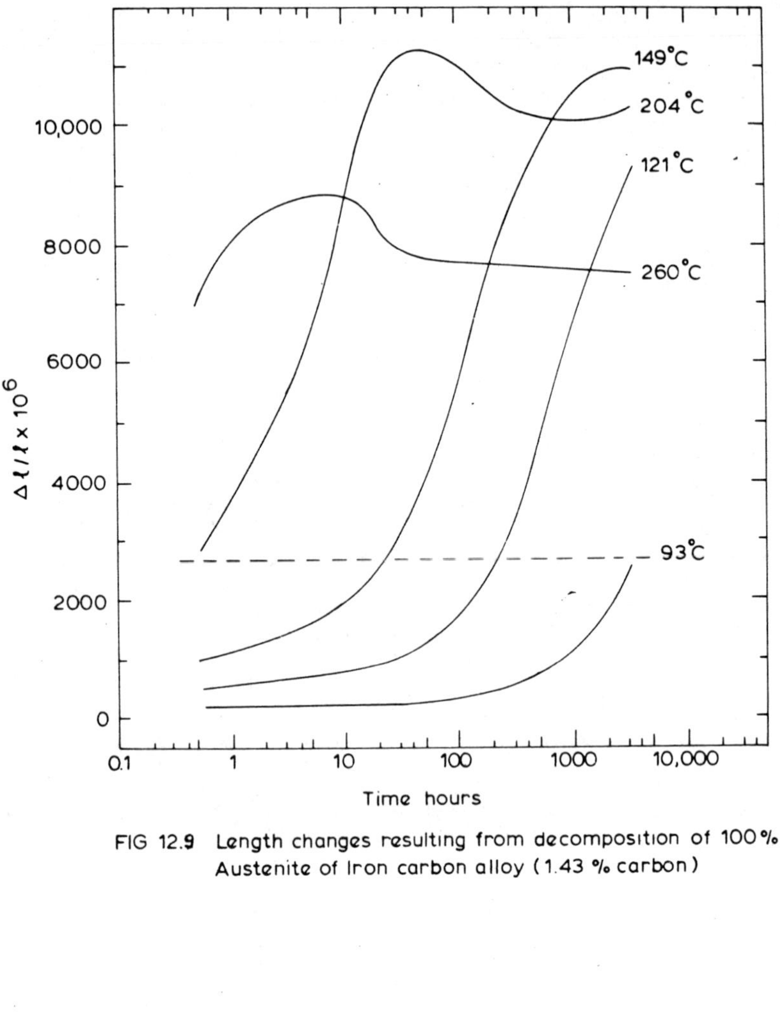

FIG 12.9 Length changes resulting from decomposition of 100% Austenite of Iron carbon alloy (1.43 % carbon)

The Arrhenius plot of the data is shown in figure 12.10

The activation energy of 103 kJ/mol may be compared with the value of 117 kJ/mol for diffusion of carbon in austenite.

Problem 12.4

Figures 12.11 and 12.12 show age hardening curves for tempering an Fe-8.0Mn alloy. Determine the activation energy for the tempering process using the times to peak hardness. To what process do you think these energies correspond?

Figure 12.13 shows the Arrhenius plot of the data.

The first ageing peak is presumably due to diffusion of interstitials to form iron carbides or nitrides. The activation energy of 111 ± 42 kJ/mol may be compared with the value of 80 kJ/mol for diffusion of interstitials in α-Fe.

The second ageing peak is due to secondary hardening to form a manganese nitride (12.15) and is controlled by diffusion of manganese in α-Fe. The activation energy of 196 ± 38 kJ/mol may be compared with the activation energy ∿260 kJ/mol for diffusion of substitutional elements in α-Fe.

References

12.1 J.M. Holloman and L.D. Jaffe, Trans AIMME, 1945, Vol 162 pp 223-249.

12.2 G.R. Speich and W.C. Leslie, Met. Trans., 1972, Vol 3, p 1043.

12.3 C. Zener, Trans AIME, 1946, Vol 167, p.550.
C. Zener, Phys. Rev. 1948, Vol 74.

12.4 W.S. Owen, E.A. Wilson and T. Bell, "High Strength Materials," John Wiley, New York, 1965, pp 167-205.

12.5 "Martensite" - Fundamentals and Technology", edited by E.R. Petty, Longman, 1970, T. Bell pp 88-91.

12.6 Ibid, F.G. Wilson, Chapter 7, pp 137-143.

12.7 F.G. Wilson and W.S. Owen, JISI, 1965, Vol 203, pp 590-596.

12.8 Y. Imai.Trans Japan Inst Metals, 1975, Vol 16, pp 721-34.

12.9 M.K. Miller, P.A. Beaven and G.D.W. Smith, "Phase Transformations", Institution of Metallurgists. April 1979, Series 3, No.11, Vol 2, pp 11-114 to 115.

12.10 G.R. Speich, Trans AIME 1969, Vol 245, pp 2553-2563.

12.11 G.V. Kurdjumov, Met Trans, 1976, Vol 7A pp 999-1011.

12.12 H.W. King and S.G. Glover, JISI, 1959, pp 123-132.

12.13 C.S. Roberts, B.L. Averbach and Morris Cohen, Trans ASM, 1953, Vol 45, pp 576-604.

12.14 C. Wells, W. Batz and R.F. Mehl, Trans AIMME, 1950, Vol 188, 1950, p.557.

12.15 B.A. Fuller and R.D. Garwood, JISI, 1972, Vol 210, p 206.

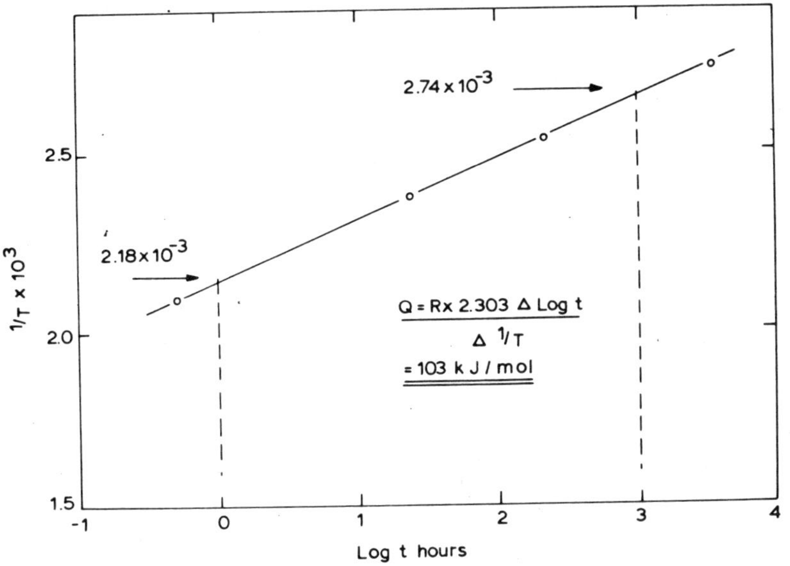

FIG. 12.10 Second-stage activation energy plot for Fe-1.43 % cabon alloy.

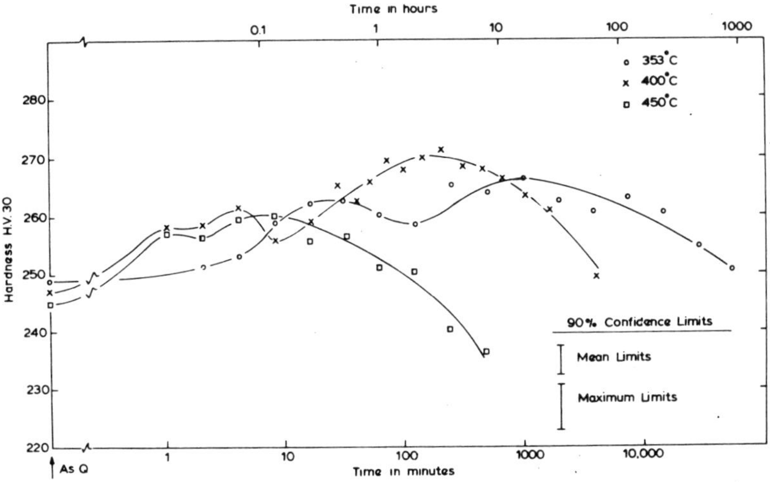

Fig. 12.11 Variation of hardness of an Fe 8.0 Mn alloy with ageing time and temperature.

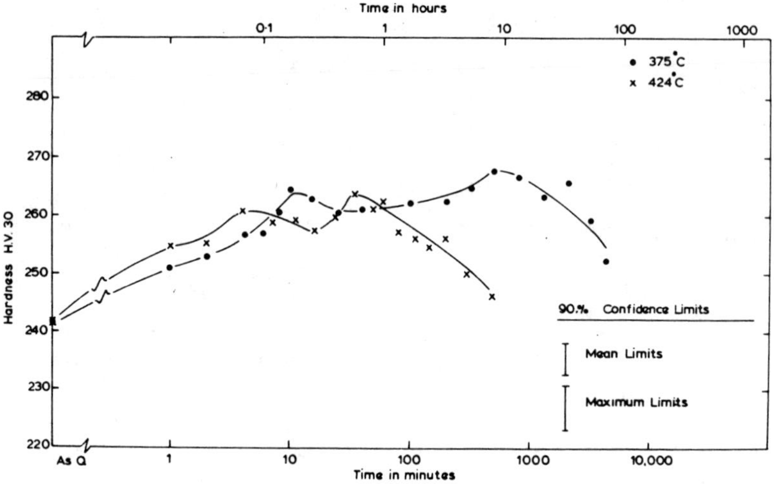

Fig. 12.12 Variation of hardness of an Fe 8.0 Mn alloy with ageing time and temperature.

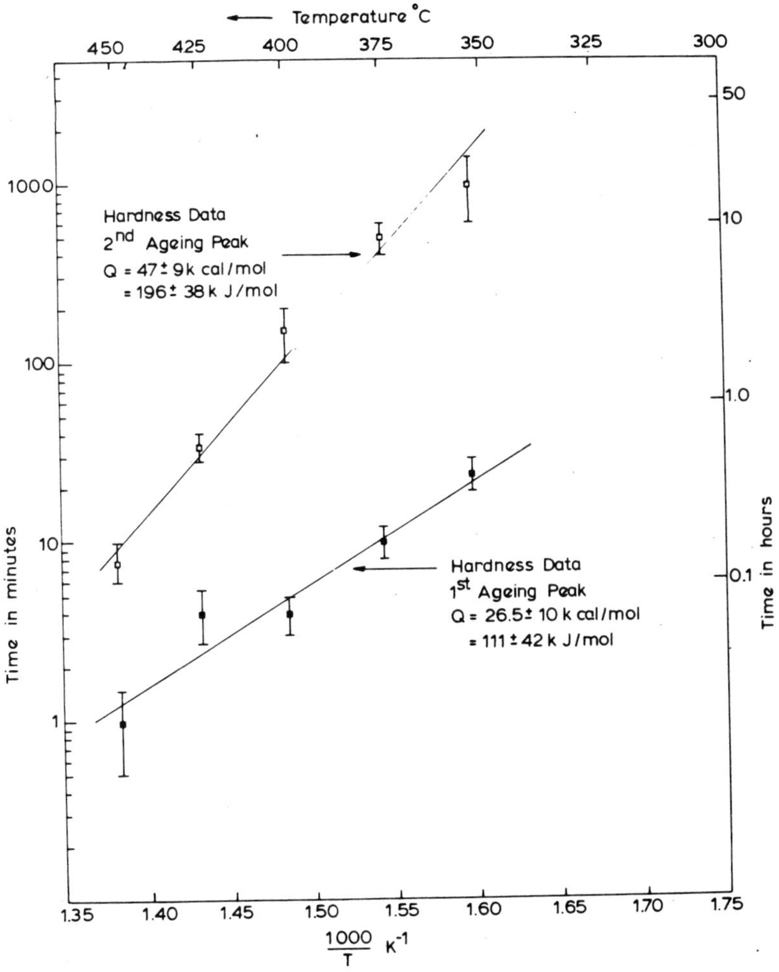

FIG. 12.13 Arrhenius plot of hardness data for an Fe-8.0 Mn alloy.

13. Coarsening of Precipitates

13.1 Introduction

On overageing precipitates tend to coarsen, ie the mean distance between precipitates increases with ageing time and large precipitates tend to grow at the expense of small ones.

The driving force for coarsening is a reduction of the total surface energy of precipitates/unit volume of alloy.

Larger precipitates will have an overall lower free energy than small precipitates, because of the reduced surface energy/unit volume.

13.2 Dependence of solid solubility on size of precipitate

The variation of free energy with precipitate size in α matrix is shown in figure 13.1.

It will be seen from this diagram, applying the common tangent rule, that smaller precipitates have a greater solubility limit N_1 in equilibrium with them, compared with larger precipitates, solubility limit N_2.

The dependence of solubility on size can be derived as follows (13.1).

Consider two spherical precipitates radius r_1 (smaller) equilibrium solubility N_1 and radius r_2 (larger) solubility limit N_2.

If we consider the activities for these particles are respectively a_1 and a_2, then we can write the reaction

$$\underset{\text{activity } a_1}{\text{small precipitates}} \longrightarrow \underset{\text{activity } a_2}{\text{large precipitates}}$$

For this reaction, equilibrium constant $K = \frac{a_1}{a_2}$ and the reduction in free energy

$$\Delta G = -RT\ln K = -RT\ln\frac{a_2}{a_1} \text{ per mole} \qquad 13.1$$

For transferring dN moles of B from particle (1) to (2)

$$\Delta G = -dNRT\ln\frac{a_2}{a_1} \qquad 13.2$$

Now the surface energy of each particle

$$S = 4\pi r^2\sigma$$

ie $\quad dS = 8\pi r dr\sigma$

$\therefore$ changes in surface energies which occur when dN moles are transferred from (1) $\rightarrow$ (2)

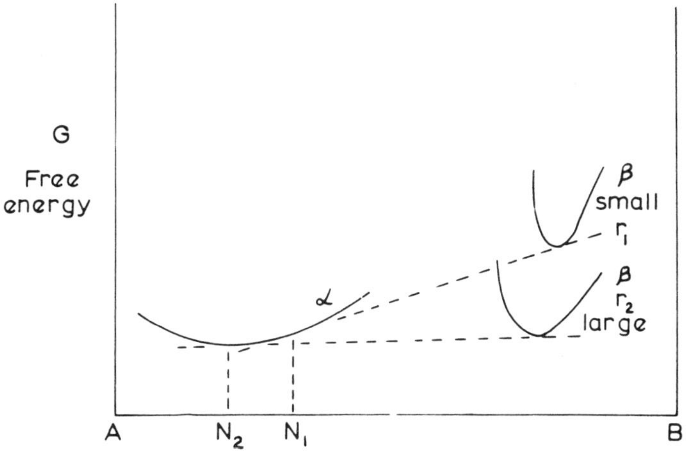

FIG. 13.1 Variation of solid solubility of α with size of precipitate β.

$$dS_1 = 8\pi r_1 dr_1 \sigma \qquad dS_2 = 8\pi r_2 dr_2 \sigma \quad \ldots\ldots\ldots\ldots\ldots \quad 13.3$$

Volume of precipitate $v = 4/3\pi r^3$

$$\frac{dv}{dr} = 4\pi r^2 \qquad \therefore\ dv = 4\pi r^2 dr$$

If Ω = volume per mol of B in precipitate, then on transferring dN moles of B from (1) → (2), change in volume = $\Omega dN = 4\pi r_1{}^2 dr_1 = 4\pi r_2{}^2 dr_2$ so that substituting in equations 13.3

$$dS_1 = 8\pi r_1 dr_1 \sigma \qquad dS_2 = 8\pi r_2 dr_2 \sigma$$

$$= 8\pi r_1 \frac{\Omega dN}{4\pi r_1{}^2}\sigma \qquad = 8\pi r_2 \frac{\Omega dN}{4\pi r_2{}^2}\sigma$$

$$= 2\left(\frac{\Omega\sigma}{r_1}\right)dN \qquad = 2\left(\frac{\Omega\sigma}{r_2}\right)dN \quad \ldots\ldots\ldots\ldots\ldots \quad 13.4$$

$\therefore$ net change in surface energy on transferring dN moles of B = $dS_1 - dS_2$

$$= \frac{2\Omega\sigma}{r_1}dN - \frac{2\Omega\sigma}{r_2}dN$$

$$= -\left(\frac{2\Omega\sigma}{r_2} - \frac{2\Omega\sigma}{r_1}\right)dN$$

which must equal ΔG on transferring dN moles of B.

$$\Delta G = -dNRT\ln K = -dNRT\ln\frac{a_2}{a_1} = -\left\{\frac{2\Omega\sigma}{r_2} - \frac{2\Omega\sigma}{r_1}\right\}dN$$

Hence $RT\ln\frac{a_2}{a_1} = 2\Omega\sigma\left\{\frac{1}{r_2} - \frac{1}{r_1}\right\}$ 13.5

If we assume an ideal solution, which is a reasonable approximation for dilute solutions

$$a_1 = N_1 \quad \text{and} \quad a_2 = N_2$$

$$RT\ln\frac{N_2}{N_1} = 2\Omega\sigma\left\{\frac{1}{r_2} - \frac{1}{r_1}\right\} \quad \ldots\ldots\ldots\ldots\ldots \quad 13.6$$

Further, if C_i = No of atoms of i/unit volume

$$N_1 = \frac{C_1}{C_1 + C_2} \qquad N_2 = \frac{C_2}{C_1 + C_2} \qquad \therefore \frac{N_1}{N_2} = \frac{C_1}{C_2}$$

$$RT\ln\left(\frac{C_2}{C_1}\right) = 2\Omega\sigma\left\{\frac{1}{r_2} - \frac{1}{r_1}\right\} \quad \ldots\ldots\ldots\ldots\ldots \quad 13.7$$

Note if $r_1 = \infty$, ie flat surface and $C_1 = C$ = solubility of particle of ∞ radius

$$RT\ln\frac{C_r}{C} = \frac{2\Omega\sigma}{r} \quad \ldots\ldots\ldots\ldots\ldots \quad 13.8$$

This is known as Gibbs-Thomson or Thomson-Freundlich equation (13.2). It can be re-written as

$$\frac{C_r}{C} = \exp \frac{2\Omega\sigma}{RTr} \quad \text{.......... 13.9}$$

Now $e^x = 1 + x + \frac{x^2}{2!} + \frac{x^3}{3!} + \frac{x^4}{4!}$

$$\therefore \quad \frac{C_r}{C} = 1 + \frac{2\Omega\sigma}{RTr} + \frac{1}{2}\left(\frac{2\Omega\sigma}{RTr}\right)^2 + \frac{1}{6}\left(\frac{2\Omega\sigma}{RTr}\right)^3 \ldots$$

Usually $2\Omega\sigma \ll RTr$ ie $\frac{2\Omega\sigma}{RTr} \ll 1$ therefore we can neglect terms higher than the first.

$$\therefore \quad \frac{C_r}{C} \simeq 1 + \frac{2\Omega\sigma}{RTr}$$

$$C_r = C\{1 + \frac{2\Omega\sigma}{RTr}\} \quad \text{.......... 13.10}$$

13.3 Diffusion controlled coarsening (13.3)

Consider a particle radius r surrounded by a large sphere radius R, figure 13.2.

Then if the particle increases in radius dr in time dt amount of material gained by particle

$$= 4\pi r^2 dr$$

Now from Fick's 1st Law the flux of atoms to particle

$$J = -D\frac{dc}{dx} = D\frac{dc}{dR} \text{ atoms/unit area/unit time}$$

ie Total flux across sphere radius R in time dt

$$= 4\pi R^2 D\frac{dc}{dR}\, dt \text{ atoms}$$

$$= 4\pi R^2 D\frac{dc}{dR}\, dt\Omega \text{ in units of volume.}$$

∴ for volume balance

$$4\pi r^2 dr = 4\pi R^2 D\frac{dc}{dR}\, dt\Omega$$

$$\therefore \quad \int_C^{Cr} D\, dc\Omega = 4\pi r^2\frac{dr}{dt}\int_\infty^r \frac{dR}{4\pi R^2}$$

$$D(Cr - C)\Omega = r^2\frac{dr}{dt}\left[-\frac{1}{R}\right]_\infty^r$$

$$= r^2\frac{dr}{dt}\left[-\frac{1}{\infty} + \frac{1}{r}\right]$$

$$= r\frac{dr}{dt} \quad \text{.......... 13.11}$$

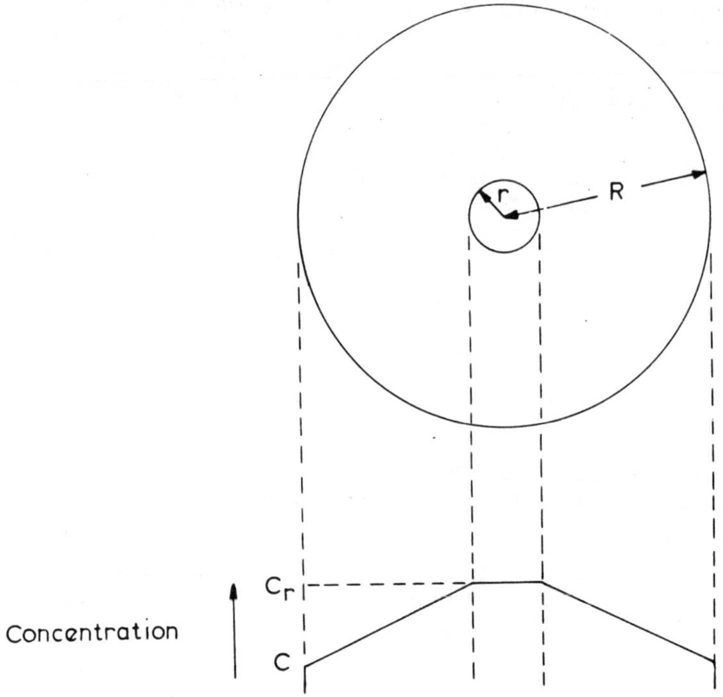

FIG.13.2 Model for diffusion controlled coarsening

But from equation 13.10

$$Cr = C\{1 + \frac{2\Omega\sigma}{RTr}\}$$

where R = Gas constant.

$$\therefore \quad Cr - C = \frac{2\Omega\sigma C}{RTr}$$

Substituting in equation 13.11

$$D\Omega \frac{2\Omega\sigma C}{RTr} = r \frac{dr}{dt}$$

ie $$dt = \frac{RT}{2\Omega^2\sigma CD} r^2 dr$$

$$\int_0^t dt = \frac{RT}{2\Omega^2\sigma CD} \int_{r_0}^{r} r^2 dr$$

$$t = \frac{RT}{2\Omega^2\sigma CD} \frac{(r^3 - r_0{}^3)}{3}$$

$$r^3 = \frac{6\sigma CD\Omega^2}{RT} t + r_0{}^3 \quad \ldots\ldots\ldots\ldots\ldots\ldots\ldots\ldots \quad 13.12$$

13.4 Interface controlled growth (13.3)

If the interface advances dr in time dt, then rate of growth of particle = $\frac{dr}{dt}$.

$\propto$ No of atoms grained by particle x frequency = $K(C_r - C)$

Substituting in equation 13.10

$$C_r - C = \frac{2\sigma\Omega C}{RTr}$$

$$\frac{dr}{dt} = K(C_r - C)$$

$$= \frac{2K\Omega\sigma C}{RTr}$$

$$\frac{RT}{2K\Omega\sigma C} \int_{r_0}^{r} rdr = \int_0^t dt$$

ie $$t = \frac{RT}{2K\Omega\sigma C} \left(\frac{r^2 - r_0{}^2}{2}\right)$$

$$= \frac{RT}{4K\Omega\sigma C} (r^2 - r_0{}^2) \quad \ldots\ldots\ldots\ldots\ldots\ldots\ldots\ldots \quad 13.13$$

Problem 13.1 BSc Hons, Final Year, Physical Metallurgy June 1977, Sheffield City Polytechnic.

The approximate Thomson-Freundlich equation, relating the concentration of solute C_r at the surface of a spherical precipitate radius r, to the matrix

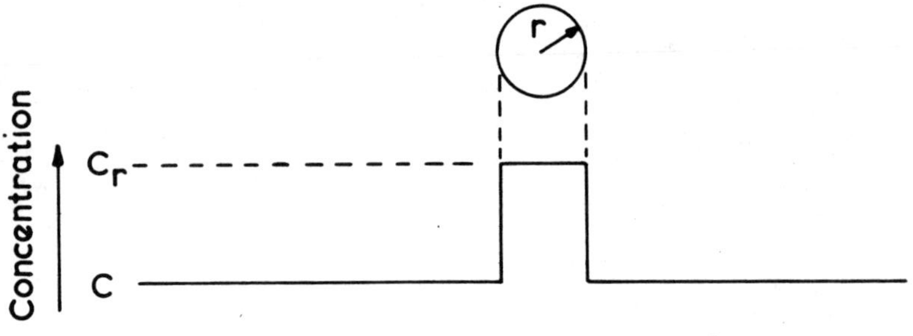

FIG. 13.3 Model for interface controlled coarsening

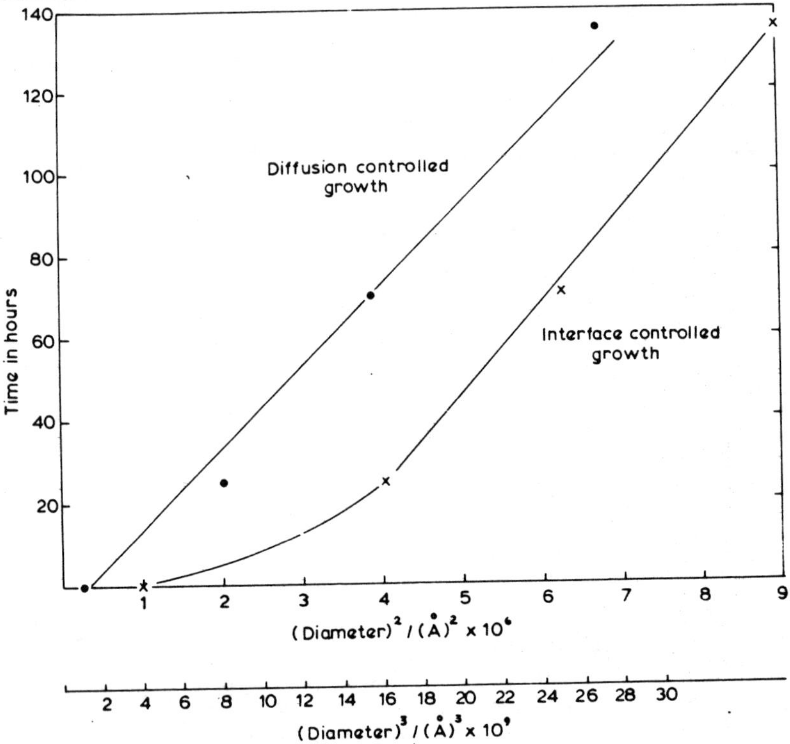

FIG. 13.4 Problem 13.1 Determination whether particle coarsening is diffusion controlled or interface controlled.

composition C is given by:

$$C_r - C = \frac{2\Omega\sigma C}{RTr}$$

where R = Gas constant

σ = surface energy of precipitate/matrix interface

Ω = volume per mole of solute

T = Temperature, K.

Show that the time dependence of precipitate coarsening is given by:

$$r^3 = \frac{6\sigma D\Omega^2 c}{RT} t + r_0{}^3 \text{ for diffusion controlled coarsening}$$

$$r^2 = \frac{4\sigma K\Omega ct}{RT} + r_0{}^2 \text{ for interface controlled coarsening}$$

where r_0 = radius of precipitate at start of coarsening, t = O

r = radius of precipitate at time t.

D = Diffusion coefficient at temperature TK.

K = temperature dependent rate constant.

The following data were obtained for the coarsening of γ^1 precipitates in Nimonic 90 at 950°C. Determine whether the coarsening is diffusion or interface controlled.

Time (hours)	0.4	25	70	135
Precipitate Diameter Å	1,000	2,000	2,500	3,000

The derivation of the equations is given in sections 13.3 and 13.4.

It will be seen from the equations given in the question that the precipitate size r^3 varies linearly with time for diffusion controlled growth and with r^2 for interface controlled growth. We will therefore plot $(\text{Diameter})^2$ and $(\text{Diameter})^3$ versus time for the precipitate to see which gives a straight line.

Time (hours)	0.4	25	70	135
Ppt. Diameter D,Å	1,000	2,000	2,500	3,000
D^2	10^6	4×10^6	6.25×10^6	9×10^6
D^3	10^9	8×10^9	15.62×10^9	27×10^9

It will be seen from figure 13.4 that the particle coarsening is diffusion controlled.

<u>Problem 13.2</u>

Given that an overageing precipitate coarsening is diffusion controlled, determine the relationship between hardness and time for an age hardening alloy.

If λ is the mean distance between precipitates then the mean shear stress τ required to bow dislocations between particles is given by (13.4),

$$\tau = \frac{Gb}{\lambda} \quad \ldots\ldots \quad 13.14$$

Assuming isotropic conditions, then yield stress,

$$\sigma_y = 2\tau \quad \ldots\ldots \quad 13.15$$

The yield stress σ_y will in turn be related to the overaged hardness increment ΔH above the as quenched or solid solution hardness

ie $$\Delta H = k\sigma_y \quad \ldots\ldots \quad 13.16$$

Combining equations 13.14, 13.15 and 13.16

$$\lambda = \frac{2Gb}{k\Delta H} \quad \ldots\ldots \quad 13.17$$

Now if it is assumed that on overageing the full volume fraction f of precipitates has precipitated and only coarsening is occurring, then quantitative metallography (13.5) gives the relationship between λ, f and the radius of particles r, as

$$\lambda = \left(\frac{2\pi}{3f} \div 2\sqrt{\frac{2}{3}}\right) r = ar \quad \ldots\ldots \quad 13.18$$

Coarsening theory gives the following relationship between coarsening time t and radius r of precipitates as

$$r^3 = \frac{6D\sigma\Omega^2 C}{RT} . t + r_0{}^3 \quad \ldots\ldots \quad 13.12$$

Hence combining equations 13.17, 13.18 and 13.12

$$\left(\frac{2Gb}{ak\Delta H}\right)^3 = \frac{6D\sigma\Omega^2 C}{RT}\, t + \left(\frac{2Gb}{ak\Delta H_o}\right)^3$$

ie $$\left(\frac{1}{\Delta H}\right)^3 = \text{constant} \times t + \left(\frac{1}{\Delta H_o}\right)^3$$

References

13.1 J. Burke, "The Kinetics of Phase Transformations in Metals", 1965, Pergammon Press, p.171.

13.2 H. Freundlich, "Kapillaschemsie, 1922, Leipzig (Akad., Verlagsgellshaft m.b.h.)

13.3 G.W. Greenwood, "The Mechanism of Phase Transformations in crystalline solids", Institute of Metals, Monograph No.33, 1969, p.103.

13.4 N.F. Mott and F.R.N. Nabarro, Report on the Strength of Solids, Physical Society, 1948, London, p.l.

13.5 R.T. Dehoff & F.N. Rhines, "Quantitative Microscopy", New York, 1968, McGraw-Hill.

14. Recovery, Recrystallisation and Grain Growth

14.1 Stored energy of cold work

Stored energy in the form of a high dislocation density provides the driving force for recovery. It is possible to calculate the stored energy of a metal from its dislocation density. The following calculation is an illustration of a simplified approach to this problem.

Problem 14.1*

Use the data given to calculate the increase in stored energy when annealed copper is heavily cold worked.

Assume:-

(i) all the stored energy is associated with dislocations;

(ii) the dislocations are randomly distributed and contain equal proportions of screw and edge components;

(iii) energy/unit length of dislocation line, E,

$$= \frac{Gb^2}{4\pi K} \ln \left(\frac{r}{r_o}\right) \qquad 14.1$$

where K = 1 for a screw dislocation

= (1-ν) for an edge dislocation;

(iv) the core radius r_o of a dislocation is 10Å.

Data:- Lattice parameter of Cu = 3.61Å; 1Å = 10^{-10}m

Shear modulus of Cu = 41.1 kN/mm^2

Poisson's ratio, ν, = 0.3

Relative atomic mass of Cu = 63.5

Density of Cu = 8.96 g/ml

Annealed dislocation density = 10^7cm^{-2}

Cold worked dislocation density = 10^{12}cm^{-2}

---

* The author is indebted to Dr B Hattersley for this problem

The crystal structure of the f.c.c. lattice is shown on the right.

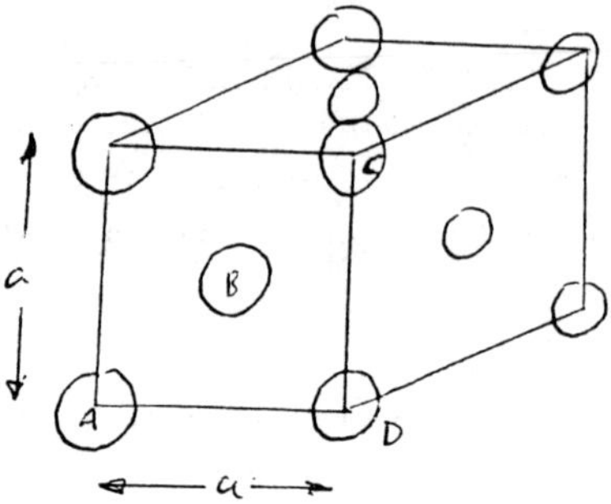

Slip occurs in the close packed direction, and therefore the Burgers vector

$b = AB$

By Pythagoras

$(AC)^2 = (AD)^2 + (DC)^2$

ie $(2b)^2 = a^2 + a^2$

$\therefore$ $b^2 = \frac{1}{2}a^2 = \frac{1}{2}(3.61)^2$

$= 6.516\mathring{A}^2$

$= 6.52 \times 10^{-20} m^2$

$= 6.52 \times 10^{-14} mm^2$

Average energy/unit length of dislocation line for both screw and edge dislocations

$$= \frac{1}{2}\left[1 + \frac{1}{(1-\nu)}\right] \frac{Gb^2}{4\pi} \ln\left(\frac{r}{r_o}\right) \quad \ldots\ldots \quad 14.2$$

$$= \frac{1}{2}\left[1 + \frac{1}{(1-0.3)}\right] \frac{41.1 \times 6.52 \times 10^{-14}}{4\pi} \ln\left(\frac{r}{r_o}\right) kN$$

$$= 2.588 \times 10^{-13} \ln\left(\frac{r}{r_o}\right) kN = 2.59 \times 10^{-13} \ln\left(\frac{r}{r_o}\right) kJ/m$$

$$= 2.59 \times 10^{-10} \ln\left(\frac{r}{r_o}\right) J/m \qquad 1J = 1Nm$$

Dislocation density = ρ line/unit area

$\therefore$ Average distance between dislocations = $1/\sqrt{\rho}$

Hence effective strain field distance, r, of each dislocation = $\frac{1}{2}\sqrt{\rho}$.

For the annealed structure $r = \frac{1}{2}\sqrt{10^7} = 1.581 \times 10^{-4} cm = 1.58 \times 10^4 \mathring{A}$.

For the cold worked structure $r = \frac{1}{2}\sqrt{10^{12}} = 5 \times 10^{-7} cms = 50\mathring{A}$.

$\therefore$ Average energy/unit length of dislocation line for annealed structure

$$E_{ann} = \frac{1}{2}\left[1 + \frac{1}{(1-\nu)}\right] \frac{Gb^2}{4\pi} \ln\left(\frac{r}{r_o}\right)$$

$$= 2.588 \times 10^{-10} \ln\left(\frac{1.58 \times 10^4}{10}\right)$$

$$= \underline{1.91 \times 10^{-9}\ J/m}$$

Dislocation density for annealed structure ρ_{ann}

$= 10^7$ lines/cm^2 $= 10^7$ cm/cm^3 $= 10^{11}$ m/m^3

∴ Stored energy/unit volume for annealed structure

$= E_{ann}\rho_{ann} = 1.906 \times 10^{-9} \times 10^{11} = \underline{1.91 \times 10^2 \text{ J/m}^3}$

∴ Average energy/unit length of dislocation line for cold worked structure, E_{cw},

$$= \tfrac{1}{2}\left[1 + \frac{1}{(1-\nu)}\right] \frac{Gb^2}{4\pi} \ln\left(\frac{r}{r_o}\right)$$

$$= 2.588 \times 10^{-10} \ln\left(\frac{50}{10}\right) = \underline{4.165 \times 10^{-10} \text{ J/m}}$$

Dislocation density for cold worked structure, ρ_{cw}

$= 10^{12}$ lines/cm^2 $= 10^{12}$ cm/cm^3 $= 10^{16}$ m/m^3

∴ Stored energy/unit volume for cold worked structure

$= E_{cw}\rho_{cw} = 4.165 \times 10^{-10} \times 10^{16} = \underline{4.165 \times 10^6 \text{ J/m}^3}$

∴ Increase in strain energy on cold working

$$\Delta E = E_{cw}\rho_{cw} - E_{ann}\rho_{ann} = 4.165 \times 10^6 - 1.91 \times 10^2$$

$$= \underline{4.16 \times 10^6 \text{ J/m}^3}$$

Note that the strain energy of the annealed state is negligible compared with that for the cold worked state.

No. of moles of Cu/unit volume

$$= \frac{8.96}{63.5} \text{ mol/ml} = \frac{8.96 \times 10^6}{63.5} \text{ mol/m}^3 \qquad 1 \text{ cm} = 10^2 \text{ m} \qquad 1 \text{ cc} = 10^6 \text{ m}^3$$

$$= \underline{1.41 \times 10^5 \text{ mol/m}^3}$$

∴ Increase in strain energy on cold working

$$= \frac{4.16 \times 10^6}{1.41 \times 10^5} = \underline{29.5 \text{ J/mol}}$$

14.2 Recovery

Recovery involves annealing out of internal strain and dislocation re-arrangement, although no visible change occurs in the optical microstructure.

The process is most easily studied by resistivity measurements although small changes in mechanical properties also occur during recovery and have been used to follow the kinetics of recovery (14.1).

It has often been found that the isothermal kinetics of recovery follow the following equation:-

$$(1-R) = x = b - a\ln t \qquad 14.3$$

$$\text{or} \quad t = \exp\frac{(b-x)}{a} = \exp\frac{(b-1+R)}{a} \qquad 14.4$$

where $R = (1-x)$ = degree of recovery, such that for no recovery $R = 0$ and $x = 1$.

t = time of isothermal holding

a & b = constants for a particular temperature.

Differentiating equation 14.3 with respect to time t, then

$$\text{Rate of recovery} = \frac{dR}{dt} = \frac{d(1-x)}{dt} = \frac{a}{t} \qquad 14.5$$

$$= A \exp - \frac{Q}{RT} \qquad 14.6$$

Then equations 14.3, 14.5 and 14.6 can be used to determine the activation energy Q for recovery from kinetic data at different temperatures.

One usually finds that a spectrum of activation energies are found depending on the value of x.

Kuhlmann (14.2) has proposed that the rate of recovery is given by the following equation:-

$$\frac{dR}{dt} = \frac{d(1-x)}{dt} = A \exp - \frac{(E+cR)}{RT} \qquad 14.7$$

This means that the activation for recovery increases linearly with R during the recovery process, which is consistent with the assumption that the most heavily deformed region of a specimen will recover first with a low activation energy.

Further Kuhlmann showed that equation 14.7 is compatible with the isothermal kinetic equations 14.3 and 14.4, since

$$\frac{dR}{dt} = \frac{d(1-x)}{dt} = \frac{a}{t} \qquad \text{(equation 14.3)}$$

$$= a \exp - \frac{(b-x)}{a}, \text{ from equation 14.4}$$

$$= a \exp - \frac{(b-1+R)}{a} \qquad 14.8$$

Thus comparing equations 14.7 and 14.8 it will be seen that Q varies linearly with R.

Problem 14.2

By following the recovery of shear stress of deformed single crystals of zinc at various temperatures (14.1), the following values of (1-R) ie x, were obtained:-

Annealing time, mins	1-R -20°C	-10°C	0°C	+10°C
5	0.856	0.789	0.656	0.504
10	0.8111	0.922	0.589	0.444
30	0.756	-	0.475	
50	-	0.578		
60	0.693	-	0.400	0.311
100	0.644	0.467	0.333	0.256

R was determined from $R = \frac{\sigma_m - \sigma}{\sigma_m - \sigma_o}$, where σ_o = initial shear stress, σ_m = deformed shear stress, σ = recovered shear stress after time t.

Show that the isothermal kinetic data obeys the equation

$$(1-R) = b - a\ln t$$

From the values of b and a determine the activation energy for recovery at R = 0.2, 0.3 and 0.5.

The kinetic data is plotted in figure 14.1, which shows that (1-R) varies linearly with logt

$$\text{ie} \quad 1-R = b - a\ln t$$
$$= b - 2.303\, a \log t$$

when $t = 1 \quad \log t = 0 \quad \therefore b = (1-R)_1$

$t = 10 \quad (1-R)_{10} = b - 2.303\, a \quad$ ie $a = \frac{b - (1-R)_{10}}{2.303}$

From figure 14.1 we obtain for a and b

Temp	$b = (1-R)_1$	$(1-R)_{10}$	$a = \frac{(1-R)_1 - (1-R)_{10}}{2.303}$
-20°C	0.985	0.822	0.0708
-10°C	0.95	0.72	0.0999
0°C	0.852	0.586	0.1155
10°C	0.63	0.45	0.4346

The activation energy, Q, for recovery can be determined from the values of 'a' as follows:

From equations 14.5 and 14.6

$$\text{Rate of recovery} = \frac{dR}{dt} = \frac{a}{t} = A \exp - \frac{Q}{RT}$$

$$\therefore \quad \ln\left(\frac{a}{t}\right) = \ln A - \frac{Q}{R} \times \frac{1}{T}$$

ie $$\frac{1}{T} = \frac{-2.303R}{Q} \log(\frac{a}{t}) + \frac{2.303A}{Q} \log A$$

cf $y = m \times x + c$

$\therefore$ A plot of $\frac{1}{T}$ versus log (a/t) is linear with slope = $\frac{-2.303R}{Q}$

(a/t) can be evaluated from values of 'a' at different temperatures TK, and the values of t corresponding to R = 0.2, 0.3 and 0.5.

Temp °C	K	1/T K^{-1}	a	t at R = 0.2	R = 0.3	R = 0.5	dR/dt = a/t R = 0.2	R = 0.3	R = 0.5
-20	253	0.003953	0.0708	13.5	55.5		0.00524	0.00128	0.00106
-10	263	0.003802	0.0999	4.5	12.2	94	0.0222	0.00819	0.0055
0	273	0.003663	0.1155	1.6	3.75	21	0.0722	0.0308	0.082
+10	283	0.003534	0.4346			53			

The Arrhenius plot of the data is shown in figure 14.2, together with the calculated values of Q.

For example, for R = 0.3

$$\text{slope of graph} = \frac{\Delta 1/T}{\Delta \log(\frac{dR}{dt})} = \frac{(3.924 - 3.71) \times 10^{-3}}{(\log 0.002 - \log 0.02)} = \frac{-2.303R}{Q}$$

$$\therefore \quad Q = \frac{2.303 \times 8.314 \times 1}{2.14 \times 10^{-4}} = \underline{89.5 \text{ kJ/mol}}$$

In actual fact when all the experimental data obtained by Drouard et al (14.1) is used an activation energy of 83.7 kJ/mol is obtained throughout the recovery process. This can be compared with the value of 85.4 ± 38 kJ/mol for self-diffusion of zinc in the c-axis direction (14.3) which is consistent with the formation of walls of dislocation of like sign and elimination of dislocations of unlike sign from such low angle boundaries.

Normally however as stated previously Q increases as R increases. (14.4).

14.3 Recrystallisation

Recrystallisation involves the replacement of the cold-worked grain structure by a set of new equi-axed recrystallised grains. The process is a kinetic one, the rate of which depends on time and temperature obeying sigmoid isothermal kinetics. The kinetics of the process also depends on the amount of prior deformation. Since increasing amounts of prior deformation increase the stored energy of the material or driving force for

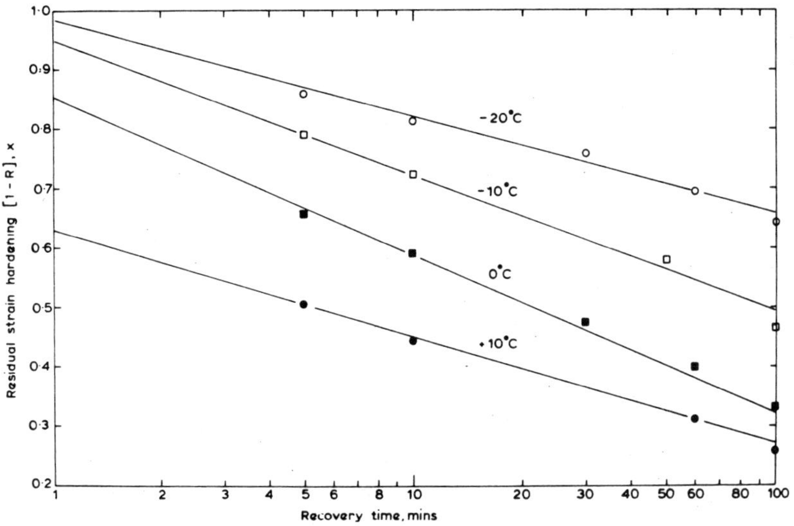

FIG. 14.1 Variation of degree of recovery of zinc single crystals with time and temperature.

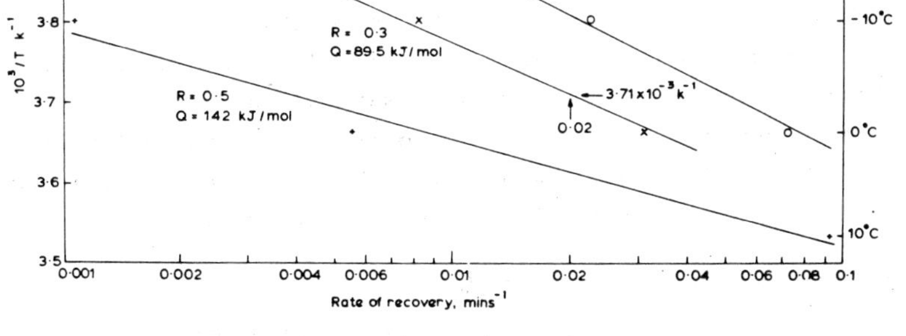

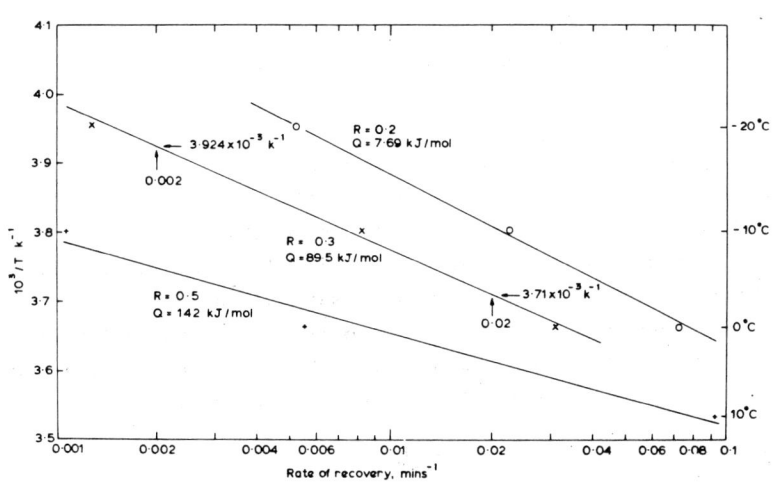

FIG. 14.2 Arrhenius plot of recovery data given in figure 14.1.

recrystallisation the rate of recrystallisation increases with the degree of prior deformation.

For a given percentage of prior deformation the rate of recrystallisation can be expressed by an Arrhenius type equation. However the activation energy obtained from such data cannot be identified with any fundamental process such as the activation energy for diffusion. This is because the processes occurring during recrystallisation are complex and involve nucleation of new recrystallised grains as well as their growth. Nevertheless a knowledge of the activation energy for recrystallisation can be useful in determining the response of a cold worked structure to an annealing treatment as illustrated in the following problems.

Problem 14.3

OFHC copper is cold worked to increase its annealed yield strength three times. It is used as a spring in a thermostat. If a safety factor of 1.5 is allowed for operation of the spring under the design stress, estimate the approximate life of the thermostat when operating at, (a) $150^{\circ}C$, (b) $175^{\circ}C$, and (c) $200^{\circ}C$.

You may assume that the time taken for 50% recrystallisation, t, is given by the Arrhenius relationship:-

$$\frac{1}{t} = A \exp - \frac{Q}{RT}$$

where $A = 10^7$ $secs^{-1}$ and the activation energy Q for 50% recrystallisation = 125 kJ/mol.

If σ = yield stress of annealed copper, then cold worked strength of the copper spring = 3σ and the operating strength = $3\sigma/1.5 = 2\sigma$. Thus when the strength of the copper has fallen to 2σ due to recrystallisation the device will fail. Further if we assume there is a linear relationship between yield stress of the copper and percentage recrystallisation, a strength of σ will correspond to 100% recrystallisation and 3σ, 0% recrystallisation and hence 2σ, 50% recrystallisation. So the device will fail when 50% recrystallisation has occurred which can be computed from the Arrhenius equation.

(a) For $T = 150^{\circ}C = 423K$, $\frac{1}{t} = A \exp - \frac{Q}{RT}$

ie Life of device, t, $= \frac{1}{A} \exp \frac{Q}{RT} = 10^{-7} \exp \frac{1.25 \times 10^5}{8.314 \times 423}$

$$= 2.731 \times 10^8 \text{ seconds}$$

$$\sim \underline{8.7 \text{ years}}$$

(b) $T = 175^\circ C = 448K\ t = 10^{-7} \exp \dfrac{1.25 \times 10^5}{8.314 \times 448}$

$$= 3.758 \times 10^7 \text{ seconds}$$

$$\sim \underline{1.2 \text{ years}}$$

(c) $T = 200^\circ C = 473K,\ t = 10^{-7} \exp \dfrac{1.25 \times 10^5}{8.314 \times 473}$

$$= 6.376 \times 10^6 \text{ seconds}$$

$$\sim \underline{74 \text{ days (ie} \sim 10 \text{ weeks)}}$$

Further recrystallisation problems involving the Arrhenius equation are given on pages 34 and 35.

The next problem illustrates the effects of prior deformation on the kinetics of recrystallisation.

Problem 14.4

Make an estimate of the annealing temperature for complete recrystallisation of iodide zirconium which has been cold rolled 51%. You may assume that the annealing time is 1h and that the activation energy for complete recrystallisation varied linearly with the percentage of prior deformation. Further you may also assume that the pre-exponential constant A in the Arrhenius equation for recrystallisation is independent of the amount of prior deformation.

Data (14.5)

For material which has received 13% prior deformation, complete recrystallisation occurs in 0.1h at 688.5°C and 1h at 620°C.

Change in activation energy for complete recrystallisation per % prior deformation = -1.06 kJ/mol, %

$R = 8.314$ J/mol K $\qquad\qquad$ $0^\circ C = 273$ K

$\log_e x = 2.303 \log_{10} x$

Working

From the Arrhenius equation

$$\text{Rate of reaction} = \frac{dy}{dt} = A' \exp - \frac{Q}{RT}$$

Assuming isokinetic behaviour

$$\frac{dy}{dt} \propto \frac{1}{t_c}$$

and $\frac{1}{t_c} = A \exp - \frac{Q}{RT}$

where t_c = time for complete recrystallisation

Q = Activation energy for complete recrystallisation

Taking logarithms

$$\ln 1 - \ln t_c = \ln A - \frac{Q}{RT}$$

ie $$\log t_c = \frac{Q}{2.303R} \times \frac{1}{T} - \log A.$$

At $T = 688.5^{\circ}C = 961.5\ K$, $1/T = 1.04 \times 10^{-3} K^{-1}$, $t_c = 0.1h$

$T = 620^{\circ}C = 893\ K$, $1/T = 1.12 \times 10^{-3} K^{-1}$, $t_c = 1h$

$$\therefore \quad Q = 2.303R \times \frac{\Delta \log t}{\Delta 1/T}$$

$$= \frac{2.303 \times 8.314\ (\log 1 - \log 0.1)}{(1.12 - 1.04) \times 10^{-3}}$$

$$= \underline{239\ kJ/mol}$$

When $t = 1h$, $\log t = 0$, $T = 893K$, $1/T = 1.12 \times 10^{-3} K^{-1}$

$$\therefore \quad \log A = \frac{Q}{2.303R} \times \frac{1}{T}$$

$$= \frac{239}{2.303 \times 8.314} \times 1.12 \times 10^{-3}$$

$$= \underline{14}$$

For 51% prior deformation

$$Q = 239 - 1.06\ (51-13)$$
$$= 199\ kJ/mol$$

$\therefore$ For 51% prior deformation and $t_c = 1h$

$$\log t_c = \frac{Q}{2.303R} \times \frac{1}{T} - \log A$$

$$\log 1 = 0 = \frac{199 \times 10^3}{2.303 \times 8.314} \times \frac{1}{T} - 13.98$$

$$\therefore \quad T = \frac{199 \times 10^3}{2.303 \times 8.314 \times 13.98}$$

$$= 743.4\ K$$

$$\sim \underline{470^{\circ}C}$$

In practice, the observed recrystallisation temperature was $567^{\circ}C$, this is because the pre-exponential constant A in the Arrhenius equation does actually vary with the amount of prior deformation.

14.4 Grain Growth

At temperatures above the "recrystallisation temperature", grain growth occurs. The driving force for this process is the reduction of the grain boundary area/unit volume, since grain boundaries possess surface energy and raise the overall energy of the metal. Boundaries migrate towards their centre of curvature. The boundaries of grains containing less than six sides are usually concave towards their centres and therefore tend to disappear, while those with greater than six sides are convex towards their centres and tend to grow.

Driving force for grain growth

Consider for simplicity a spherical grain of initial radius R which shrinks by an increment dR due to boundary migration. Suppose that dN atoms cross the boundary during this process.

Then the free energy change accompanying this transfer

$$= \Delta\mu . dN = \sigma dA \quad \text{......} \quad 14.9$$

where $\Delta\mu$ = change in chemical potential/atom

σ = grain boundary energy

dA = change in grain boundary area.

For a sphere $V = 4/3\pi R^3 \quad \therefore \; dV/dR = 4\pi R^2$ 14.10

$A = 4\pi R^2 \qquad dA/dR = 8\pi R$ 14.11

If Ω = atomic volume = volume occupied by one atom, then

$$dV = \Omega dN \quad \text{......} \quad 14.12$$

Hence, substituting equations 14.12 in equation 14.9 gives

$$\Delta\mu = \sigma \frac{dA}{dN} = \sigma\Omega \frac{dA}{dV} \quad \text{(from equation 14.12)}$$

$$= \frac{2\sigma\Omega}{R} \quad \text{(from equations 14.10 and 14.11)}$$

If we assume that $R \propto D$ where D = mean grain diameter

ie $R = \beta D$ where $\beta \simeq 0.5 - 1.0$

Taking $R = \frac{D}{2}$ for a spherical grain

Then $\Delta\mu = \frac{4\sigma\Omega}{D}$ 14.13

Thus the driving force for grain growth is inversely proportional to the grain diameter.

Rate of Grain Growth

It is reasonable to assume that rate of grain growth is proportional to the driving force which in turn is inversely proportional to the grain diameter.

ie $$\frac{dD}{dt} = \frac{k}{D} \quad \ldots\ldots \quad 14.14$$

Integrating $D^2 = kt + \text{Constant}$

when $t = 0$ at start of grain growth, $D = D_o$

$$\therefore \quad D^2 - D_o^2 = kt \quad \ldots\ldots \quad 14.15$$

This equation is obeyed by pure metals and some single phase alloys.

The effect of temperature is given by the Arrhenius equation

$$k = A \exp - \frac{Q}{RT} \quad \ldots\ldots \quad 14.16$$

More generally a growth law of the form

$$D = (kt)^{n}. \quad \ldots\ldots \quad 14.17$$

is found with $n < 0.5$. This is attributed to impurities exerting a drag on the moving grain boundaries.

Problem 14.5

The following grain growth data were obtained for alpha-brass (14.6).

Annealing Time (mins)	Grain Diameter, μm			
	700°C	650°C	600°C	550°C
30	45	40	26	19
60	65	53	40	25
90	80	65	51	28
120	90	73	58	31

Show that the data obeys the ideal grain growth law and determine the activation energy for grain boundary movement.

If we assume that D_o is small compared with D, then we can write

$$D = (kt)^{n}.$$

ie $\log D = n \log t + n \log k$

cf $y = mx + c$

Thus a plot of log D versus log t should be linear of slope $n = \frac{1}{2}$ for ideal grain growth.

The data given in the problem is plotted in figure 14.3. It will be seen from this figure that only the data determined at 700°C are close to the ideal grain growth law. This is because we have not allowed for the initial grain size D_o. The grain diameter data for 700°C varies over a wide range and therefore only in this case can we neglect D_o.

A more accurate verification of the ideal grain growth law is obtained by plotting D^2 versus time, t, figure 14.4.

Now the slope of the lines in figure 14.4 gives values of the rate constant k at each temperature.

Hence an Arrhenius plot of k, figure 14.5, enables the activation energy for grain growth to be determined.

Problem 14.6

The following data were obtained during the isothermal annealing of a cold worked metal at the temperatures shown.

Annealing time/h	Mean Grain Diameter/μm	
	600°C	700°C
0.50	30	47
0.75	36	57
1.25	46	73
2.00	58	91

(i) Determine the grain growth exponent and comment upon its value.

(ii) Calculate the activation energy for grain growth.

Gas constant R = 8.314 J/molK.

Limiting grain size due to second phase particles

When a dispersion of second phase particles is present in an alloy, grain growth is severely inhibited, typically $n \sim 0.02$. The degree of grain refinement achieved depends on the volume fraction f of particles present and the size of the particles. In general the smaller the size of the particles and greater the volume fraction, the smaller the grain size.

We will now derive a relationship between the grain size D, volume fraction f, radius of particle r for spherical grains and spherical particles. This relationship is due to Zener (14.7).

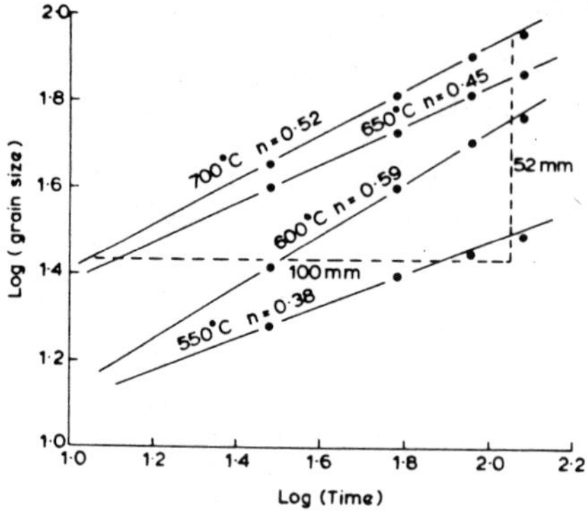

FIG. 14.3

Kinetic analysis of grain growth in alpha brass.

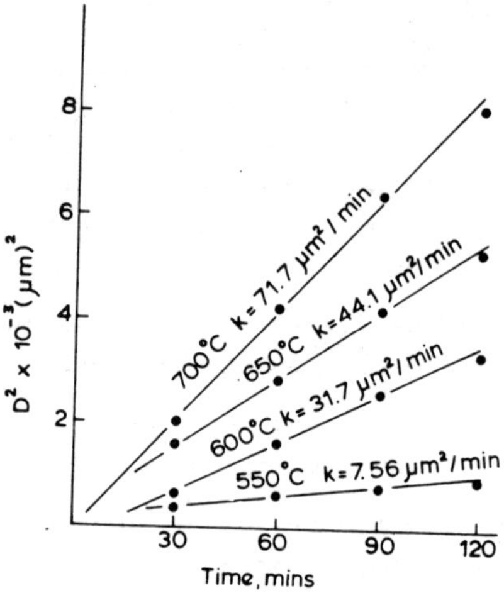

FIG. 14.4 Kinetic analysis of grain growth in alpha brass using equation $D^2 - D_0^2 = kt$ where D = grain diameter

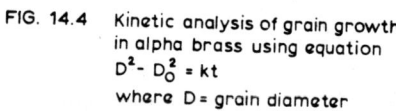

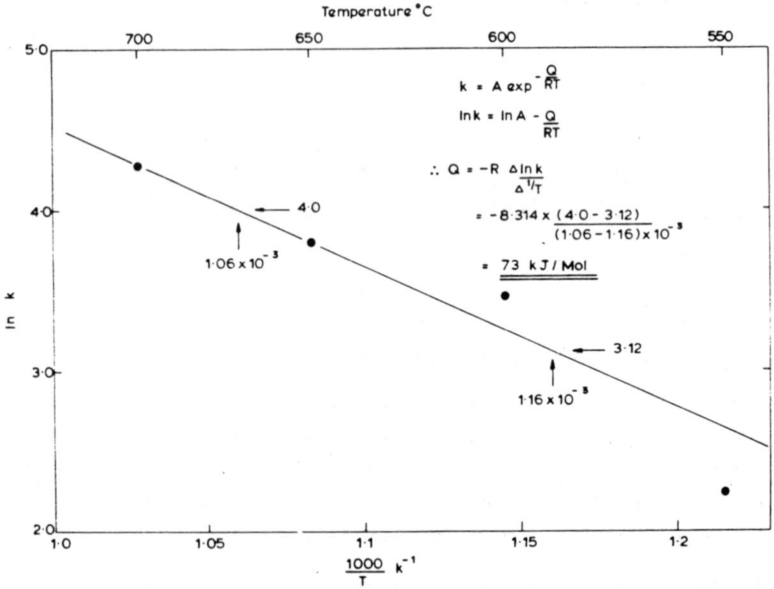

FIG. 14.5 Arrhenius plot of kinetic data for grain growth in alpha brass, taken from figure 14.4

Drag force of a particle on a boundary

Consider a spherical particle radius r restraining the movement of a grain boundary, figure 14.6.

As the boundary tries to move away from the particle, new grain boundary area has to be created and this acts as a drag on the movement of the boundary. This additional grain boundary area holds the boundary back until the surface energy component acting perpendicular to the boundary is sufficient to detach the boundary from the particle. The maximum value of the component is equal to the drag exerted by the particle on the boundary.

In figure 14.7, resolving perpendicular to the boundary

Drag force/particle F_p

= Length of grain boundary in contact with particle x surface energy component perpendicular to boundary

$= 2\pi r^2 \times \sigma \sin\theta$

$= 2\pi r \cos\theta \times \sigma \sin\theta$

$= \pi r \sigma \sin 2\theta \quad (\sin 2\theta = 2\sin\theta \cos\theta)$

This is a maximum when $\theta = 45^\circ$

$\therefore \quad F_{max} = \pi r \sigma$/particle 14.18

Let there be n particles/unit volume.

Then volume fraction of particles $f = \frac{4}{3} \pi r^3 n$

ie $\quad n = \frac{3f}{4\pi r^3}$ 14.19

If we assume a uniform distribution of particles, then the boundary will intersect particles whose centres lie within +r to -r either side of boundary, see figure 14.6.

Hence unit area of the boundary intersects those particles within a volume 2r x 1.

There are n particles/unit volume, therefore there are 2r x 1 x n = 2rn particles/unit area of grain boundary.

$\therefore$ Maximum drag force/unit area of boundary F_A = drag force/particle x number of particles/unit area of boundary

$= \pi r \sigma \times 2rn = \pi r^2 n \sigma$

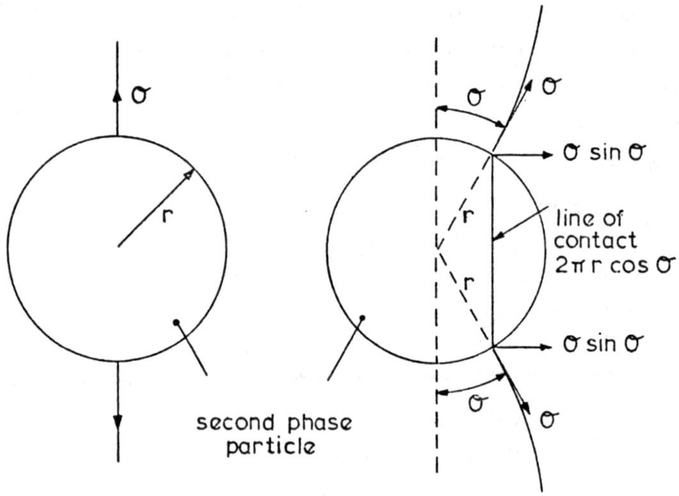

FIG. 14.6 & 14.7 Second phase particle restraining movement of a grain boundary (14.11)

Substituting in $n = \frac{3f}{4\pi r^3}$, equation 14.19

$$F_A = \frac{3f\sigma}{2r} \quad \text{.......................................} \quad 14.20$$

Consider unit area of grain boundary migrating unit distance

Then work done $= F_A \times 1$

= energy released/unit volume

= change in chemical potential/atom x Number of atoms/unit volume

$= \Delta\mu \times \frac{1}{\Omega}$

But from equation 14.13, $\Delta\mu = \frac{\sigma\Omega}{D}$

$\therefore \quad F_A = \frac{\Delta\mu}{\Omega} = \frac{\Delta\sigma}{D}$

$= \frac{3f\sigma}{2r}$ from equation 14.20.

$$\text{Hence } D = \frac{8r}{3f} \quad \text{.......................................} \quad 14.21$$

Problem 14.7

In HSLA steel, grain refined with 0.03% Nb the following data was observed at 950°C

Radius of austenite grains = 14μm

Mean size of NbC particles = 250Å

Check the consistency of the data with Zener's equation 14.21. You may assume that all the NbC is precipitated as fine carbides.

Relative atomic mass of Nb = 92.91

Relative atomic mass of C = 12.01

Density of NbC = 8.2 g/ml

Density of Fe = 7.9 g/ml

We must first of all convert the mass fraction of Nb into volume fraction of precipitate f.

92.91 grams of Nb combines with 12.01 grams of C to form 104.92 grams of NbC.

$\therefore$ 0.03 grams of Nb forms $\frac{0.03}{92.91} \times 104.92$ grams of NbC

$\therefore$ Volume fraction of NbC $f \simeq \frac{0.03}{92.91} \times 104.92 \times \frac{8.2}{7.9}$

$\simeq 0.035\%$ volume

$= 3.5 \times 10^{-2}/10^2 = 3.5 \times 10^{-4}$ volume fraction

∴ Smallest grain size which can be effectively pinned by the dispersion $D = {}^{8r}/_{3f}$

$$= \frac{8 \times 125}{3 \times 3.5 \times 10^{-4}} \text{Å} \qquad r = \tfrac{1}{2} \times 250\text{Å}$$

$$= 9.5 \times 10^{5}\text{Å} = 9.5 \times 10^{5} \times 10^{-10}\text{m} = \underline{95\ \mu\text{m}}$$

This overestimates the austenite grain size because Zener's equation considers only one grain and does not allow for the interaction of grain surface; ie the decrease in size of one grain involves an expansion of another.

Hillert (14.8) was the first to allow for this and a sophisticated treatment by Gladman (14.9) gives the critical particle radius r* for pinning grains as

$$r^* = \frac{6R_o f}{\pi} \left(\frac{3}{2} - \frac{2}{Z}\right)^{-1}$$

where f = volume fraction of precipitate

R_o = matrix grain size = ½ x diameter

Z = ratio of the radii of the growing grain and its neighbour.

References

14.1 R. Drouard, J. Washburn and Earl R. Parker, Trans AIME., 1953, vol. 197, pp 1226-1229.

14.2 Kuhlmann-Wilsdorf, D., Z Physik, 1948, Vol.124, p 468.

14.3 P.H. Miller and R.F. Banks, Physical Review 1942, Vol.61, p.648.

14.4 J.T. Michalak and H.W. Paxton, Trans AIME 1961, Vol.221, pp 850-857.

14.5 R.M. Treco, Proc. AIME Regional Conference on Reactive Metal. 1956, p.136.

14.6 P. Feltham and G.J. Copley, Acta Met., 1958, Vol.6, p.539.

14.7 Quoted in C.S. Smith, Trans AIME, 1948, Vol 175, p.15.

14.8 M. Hillert, Acta. Met., 1965, Vol 13, p 227.

14.9 T. Gladman, Proc. Roy. Soc., 1966, (A), Vol 294, p.298.

General References

14.10 R.W.K. Honeycombe, "The Plastic Deformation of Metals", Edward Arnold Ltd., 1968, pp. 283-321.

14.11 Robert E. Read-Hill, "Physical Metallurgy Principles". D. Van Nostrand, Inc, 1964, pp.175-219.

15. Answers

Problem 1.2 $\Delta S^{o} = \underline{\underline{-69\ J/mol}}$ $\Delta H^{o} = \underline{\underline{44\ kJ/mol}}$

Problem 3.3 Graphically for t - 1h 1/T = 0.001504

$T = 665\ K = \underline{\underline{392^{o}C}}$

Problem 3.4 $x = \underline{\underline{2.08\ mms}}$

Problem 3.5 Upper bainite Q = $\underline{\underline{136\ kJ/mol}}$ should correspond to diffusion of carbon in austenite.

Lower bainite Q = $\underline{\underline{29.5\ kJ/mol}}$ should correspond to diffusion of carbon in ferrite.

Problem 4.3 $399^{o}C$ $n = \underline{\underline{2.0}}$

$269^{o}C$ $n = \underline{\underline{3.6}}$

The values of n are dissimilar since the transformations are not the same; at $399^{o}C$ upper bainite is forming while below $350^{o}C$ at $269^{o}C$ lower bainite is forming.

Problem 4.4 Values corresponding to 75%, 80%, 85%, 90% and 95% transformation may be obtained by plotting log log $(\frac{1}{1-y})$ versus log t,or on log/log graph paper, log $(\frac{1}{1-y})$ versus t and extrapolating the straight line obtained to higher values of y.

75% 840 secs; 80% 880 secs; 85% 920 secs; 90% 970 secs; 95% 1050 secs.

The reaction is slowing down due to exhaustion of nucleation sites and impingement.

Problem 5.4 $\delta = 0.0056$ is the same in all directions so precipitate grows as a sphere

$$r^* = \frac{-2\sigma}{(\Delta G_v + W)} = 3.47 \times 10^{-6}\ mms$$

$$= \underline{34.7\mathring{A}}$$

where $\Delta G_v = \frac{\Delta H_v \Delta T}{T_o}$ and $W = 6G\delta^2$